AF859451

PERSPECTIVES ON APPLIED PHYSICAL GEOGRAPHY

by

John E. Oliver
Indiana State University

Stephen L. Stover
Kansas State University

M. Duane Nellis
Kansas State University

and

Charles E. Bussing
Kansas State University

KENDALL/HUNT PUBLISHING COMPANY
2460 Kerper Boulevard P.O. Box 539 Dubuque, Iowa 52004-0539

This edition has been printed directly from camera-ready copy.

ISBN 0-8403- 5812-1

Printed in the United States of America
10 9 8 7 6 5 4 3 2 1

TO OUR STUDENTS: Past, Present, and Future.

Contents

Part II The Atmosphere

Preface to the Second Edition

When the first edition of this book went out of print and thus became unavailable to our students in environmental geography, three of us at Kansas State University set out to replace it. The result was an agreement to join John Oliver, the original author, in a revision that would again assist our students, as well as others, to perceive the practical implications of textbook theory and process.

The book's basic arrangement is unchanged, but all of the surviving original chapters have been updated or re-written. The chapter on solar energy has been replaced by a completely new one, located in the "Biosphere" section. "Aerosols, Ozones, and Life" has been supplanted by "Acid Rain", and a completely new and engrossing essay on climatic change replaces the earlier one with the same title. With reluctance we have deleted the chapters on weather modification, earthquakes, scenery, trans-Amazon highway, and the earth's productivity. In their places we have substituted a timely look at the rainforest's future especially in Amazonia, and a chapter on severe storms. Finally, we have added a concluding chapter, "The Human Perspective", where one of today's most perceptive writers offers a summary view of our environmental future.

Like most projects of its kind, this revision has taken much longer than we had anticipated, and we acknowledge with thanks the patience of our publisher and our respective families, as well as the help of our able secretaries, their assistants, and our cartographers at Indiana State University, Kansas University, and Kansas State University. We also appreciate those who have helped with photocopying and those who have generously provided illustrative and other material for this revision. However, along with whatever credit may accompany this updated and enlarged book of readings in physical geography, we must also stand ready to accept responsibility for errors that may have eluded our best efforts to detect and remove them.

J.E.O.
C.E.B.
M.D.N.
S.L.S.

Preface to the First Edition

Relatively few of the students who take introductory physical geography courses will ever become professional geographers. In most cases the single course represents their only exposure to the subject. For this reason the course must lay a basis for much of the understanding of the physical world--and not solely provide technical information to be memorized. It is an important course, one that requires much thought and input from the instructor. I hope that this book will help the instructor introduce students to physical geography as it applies to the world they know--both in their immediate experience and on a wider scale.

It is my purpose in this book to show that principles of physical geography apply to actual situations and that they play a primary role in the everyday affairs of people of the world over. Introductory physical geography texts cannot possibly include all processes and their applications. Something has to be sacrificed and it is generally the applications, because the main objective of the course is the presentation of fundamental concepts so that students can understand the major features of our physical world. Additionally, the physical geographer's concern with ongoing environmental changes means that, as research continues, the base of physical geography continues to expand. As a result, the instructor, limited by time considerations, and the textbook authors, limited by space, must place inordinate emphasis on processes and models exemplifying the existing conditions. Through necessity, the real-world meaning of the physical principles covered is relegated to a secondary position.

The need to supplement basic concepts with reading material in which real world situations are used is recognized by many teachers, and "outside readings" have long been an integral part of many physical geography courses. Unfortunately, much work that is interesting and informative is published in research journals that make very difficult reading for the novice. In an effort to meet this problem, the chapters in this book draw upon the research of many workers and present their findings at a level appropriate to the introductory student. Rather than providing a set of prepublished works, which often fail to convey the exact message, I present a series of original essays.

The contents of the essays are based upon my experiences in teaching physical geography and environmental science over a number of years. Selection of the topics, which I believe represent a cross section of material typically dealt with, was one of the most difficult tasks in preparing the book. With page limitations, it was necessary to omit many interesting applications of major importance to the physical geographer. I trust, however, that the rationale for including the topics presented becomes apparent. Some essays, such as the chapter on map projections, attempt to give meaning to material that seldom seems to generate much interest in students. Others, such as the account of potential ozone depletion, focus attention on subject matter that is currently found either in highly technical research reports or in vague accounts in popular magazines and newspapers.

The nature of the problems covered makes this book potentially useful in a variety of environmental courses, in addition to the obvious ones in physical geography. For either purpose a number of items have been included to facilitate the book's use. Each chapter is preceded by a Context page that places the given study in the context of formal textbook and classroom coverage. At the end of each chapter there is a list of suggestions for further work. I have offered various approaches in the questions because of the difficulties involved in completing additional work in large introductory courses. Perhaps some of the approaches could be used as discussion topics, as an alternative to the completion of set exercises in "laboratory" sessions. Others could be answered, at least in part, by careful reading of the selected textbook.

The very nature of the content of the essays required the use of the works of many individuals. I trust that in presenting their efforts at an appropriate level, I have succeeded in retaining the correct interpretation of their findings. In acknowledgement of those works I have cited the appropriate sources at the ends of the chapters. Many other persons also contributed to the preparation of this book. Particular thanks go to the reviewers of the manuscript and to those who took time to contribute additional material to their published work: John Emerson, Helmet Lieth, Marie Morisawa, Paul Mausel, Alan Strahler, and Leo Weisbecker. I am grateful to my colleagues at Indiana State University for the time spent in discussing many of the topics. Finally, for her considerable patience and ready acceptance of a number of postponed vacations, much gratitude goes to my wife.

J.E.O.

Introduction

Physical geography is a very relevant area of study in today's world. No matter how technical the subject seems, we can find material of great interest concerning the interrelationships between human beings and their physical environment. The tremendously broad scope of physical geography is apparent in the contents of the many introductory textbooks available. Topics range from the explanation of day-to-day weather to the discussion of the ways in which rivers and glaciers modify the earth's surface. Such an examination provides a scientific explanation of many of the environmental processes that we observe continuously around us.

The importance of this book in the study of physical geography is that it adopts an applied rather than a theoretical approach. Each chapter takes a basic physical geographic idea or process and shows how it actually applies to the everyday lives of people. Even theoretical concepts have a ready application over wide areas of the world. For example, the patterns of the general circulation of the atmosphere can be related to the recent drought and famine suffered by people in West Africa. At a more local level, the theories of slope stability are of importance to skiers and owners of mountain homes.

In some cases the applied approach relies upon examination of the historical development of a geographic fact. Many of the items we now accept as facts had first to be proven, and herein lie many fascinating tales of peoples' endeavors to understand the physical world. Who, for example, first measured the size of the world, and what were the consequences of the measurement? Who developed the idea of drifting continents, and how does that early idea relate to present-day interpretations of earthquakes? By examining some of the scientific theories behind such developments we can begin to understand some of the puzzling and important aspects of the world.

While the cases we shall explore essentially concern physical geographic principles, they also touch upon other branches of geography. In looking at the assets and liabilities of the physical world we must remember that these factors concern people: their beliefs, their cultural background, and their level of technology. Can we, for instance, criticize the use of water resources in Pakistan without understanding the background of the people concerned? Is it possible to evaluate medieval beliefs about the shape of the world without an awareness of the state of knowledge at that time? Studies in applied physical geography need to consider carefully the social, cultural, and historical situation before coming to any conclusion.

In sum, the purpose of this book is to help you, the student, to relate the concepts of physical geography to the world that you experience. The relationship between physical processes and technical, economic, and social developments of the modern world is often uncomfortable and sometimes conflicting. As a citizen you will be called upon to make decisions involving these relationships. We hope that reading this book will help you formulate sounder and better informed judgments than otherwise might be the case.

Part I
Views of the Earth

Chapter 1

Earth Facts:
The Size and Shape of the Earth

The Context

The first photographs of the earth taken from space created a sense of awe and excitement: here, at last, was a space-age view of the earth! Through our increasing familiarity with views such as the figure on the next page from books, articles, and even advertisements we have lost that initial excitement. Instead, we see such photographs merely as an illustration of fact: the

Figure 1-1. Planet Earth. (Photo courtesy of U.S. National Aeronautics and Space Administration).

earth is a spherical planet with a bluish tinge. It may well be that for future generations the story of the development of the space program will be only another subject in history books.

Today most of the important discoveries about the earth are simply history for us. In studies of the earth and its environment, basic dimensions are typically as follows:

equatorial radius:	3,963.5 miles (6,377.2 kilometers)
polar radius:	3,950.2 miles (6,357.0 kilometers)
circumference at equator:	24,900 miles (40,071.6 kilometers)
weight (mass):	6.6×10^{21} tons (5.97×10^{27} grams)

It is hard for students to get excited about such statistics. Questions like: "When were the basic facts discovered?" "Who discovered them?" or "What were the prevailing ideas before the facts became facts?" are rarely asked. Yet, like the development of the space program there is a fascinating story of discovery behind each dry fact. In this chapter we shall examine the basic ideas about the size and shape of the earth from a historical standpoint.

Today the Flat Earth Society, with its headquarters in Dover, England, does not have very many members. This is not surprising, for most people have been taught since childhood that the earth is a sphere. But what if we had not been exposed to the scientific facts that prove that the earth is round? Suppose, for example, you were a Bambuti pygmy. Your world would be a rainforest. The sun would not be a bright disc crossing the sky but a series of flickering patches of light. The stars would barely, if ever, be visible. The dense, luxuriant forest growth would destroy a sense of distance, and all things would appear at close range. In such an environment is there any evidence of a round earth?

Although the Bambutis' limited view of the horizon contributes toward their idea of earth shape, such a confining environment is not necessary for a person to have a view of the earth as flat. In his study of human perception of the world, geographer Yi-Fu Tuan writes: "The idea of the earth as a flat disc surrounded by water and floating on it appears in many parts of the world. The Yurok Indians of northern California, for example, appear to conceptualize their world two-dimensionally as a circular disc, despite the ruggedness of their home ground" [5:35-36]. The map of the world drawn by the

Yurok Indians (Figure 1-2) shows the Klamath River, on which the Indians depend for their livelihood, as the main geographic feature. It bisects their world. The points of the compass are oriented by the river location, and somewhere on the banks of the Klamath is the center of the world.

Figure 1-2. The world as seen by the Yurok Indians (From Yi-Fu Tuan, Topophilia: A Study of Environmental Perception, Attitudes, and Values, Copyright 1974, p. 36. Reprinted by permission of Yi-Fu Tuan, Department of Geography, University of Wisconsin, Madison).

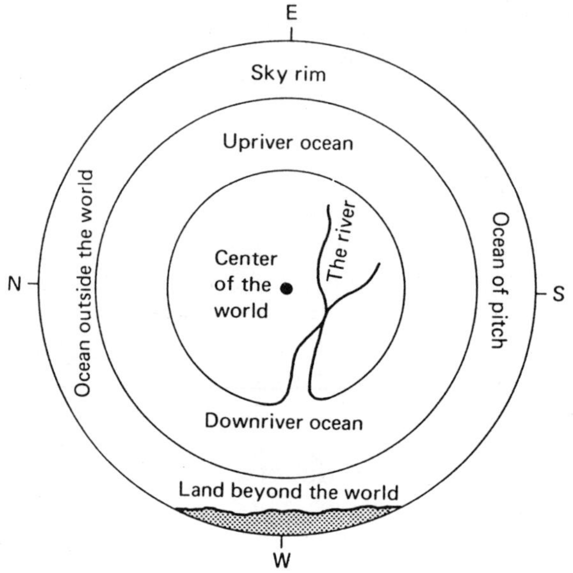

Many persons believe today that Columbus set out in 1492 with a unique vision of a round earth. This certainly was not the case. While unschooled people living in the fifteenth century may have had no notion of the real shape of the world, informed scholars had a great deal of knowledge to draw on. In fact, their basic problem was not imagining a round earth but rather estimating the distance around it.

DATA FROM ANCIENT GREECE

If the fifteenth-century European explorers had only maps made in Europe during medieval times to draw upon, Columbus might never have set sail. Medieval times were bleak so far as cartography was concerned. A typical representation of the earth was the T-in-O map, a circular map in which an ocean sea surrounded land that was itself divided into three parts by bodies of water. As we can see in Figure 1-3, Asia occupies the upper half of the circle and is bounded by the rivers Don and Nile, and Europe and Africa are divided by the Mediterranean Sea. The pattern of land and water makes a T-shape that is surrounded by the "O" of the ocean sea. Such maps, drawn by theological scholars, served no geographical purpose but were symbolic representations of the world from a religious point of view. Following Christian theology, the scholars placed Jerusalem at the center of the world and, in keeping with the Trinity, divided the world into three parts. In some of the maps they emphasized the biblical significance further by identifying the three land divisions as the lands given to Noah's sons: Asia was called Shem; Europe, Jafeth; and Africa, the land of Ham.

However, territorial aggresion brought new, more realistic maps to Europe. When the Turks expanded westward in the fourteenth century, people fled from Byzantium - the modern city of Istanbul. Naturally, they took treasures with them, and among these were texts by Ptolemy, an Alexandrian astronomer of the second century AD. Ptolemy's manuscripts were essential in understanding the dimensions of the world.

Ptolemy's work, Geographia, contained much geographic information and presented many of the philosophical ideas of the earlier Greek scholars. One of the most important was the round earth concept, proposed in the fifth century BC by Pythagoras and his followers and later adopted by Platonic philosophers. In accordance with the philosophy of a round earth Ptolemy drew up instructions for making map projections

Figure 1-3. The medieval theologians' symbolic representation of the world depicted by the T-in-O map. (Redrawn from Raisz, General Cartography (New York: McGraw-Hill, 1948.) By permission.)

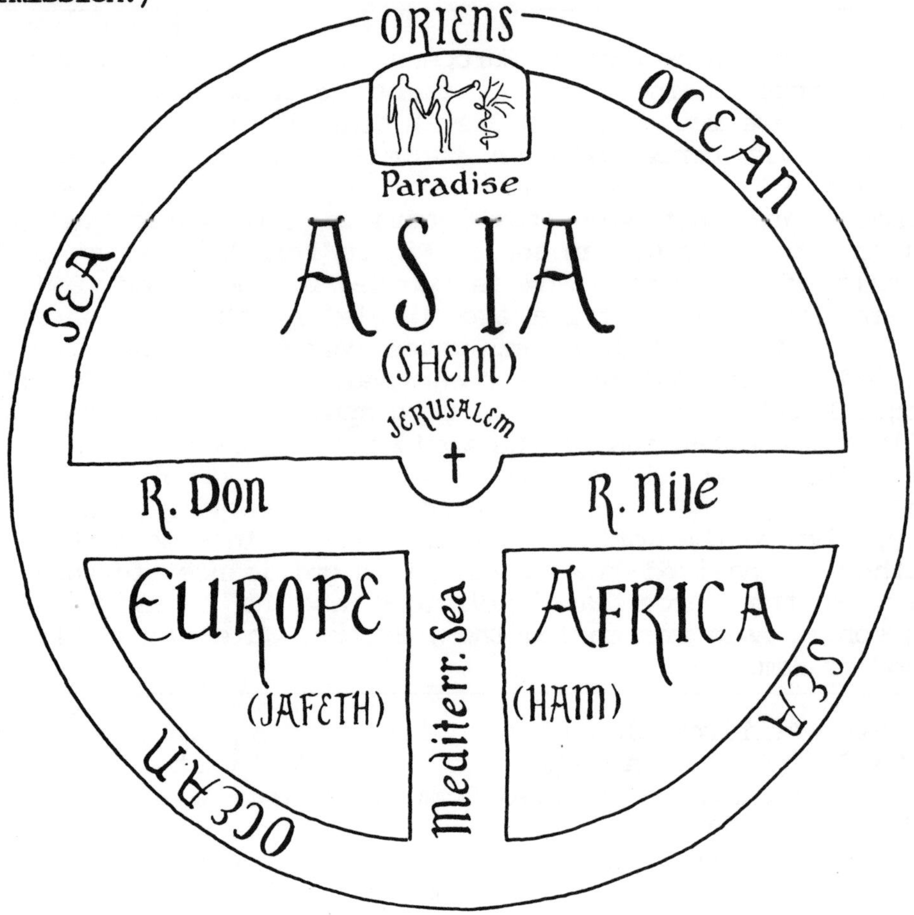

and listed coordinates for 8,000 places. These places were given as latitude and longitude in degrees, the latitude based upon relative length of day and the longitude on time, with one hour the equivalent of fifteen degrees.

In drawing upon the work of earlier scholars, Ptolemy made one error that had important consequences for the fifteenth-century voyages of discovery. As has been described in many introductory textbooks, the Greek astronomer, Eratosthenes (276-196 BC), had calculated the size of the earth by measuring the distance between Alexandria and Syene and by observing the size of shadows cast by the overhead sun at those locations. Using simple geometry, he calculated the earth's circumference as 250,000 stades, a value that when converted to current units is only about 320 kilometers (200 miles) short of the actual circumference. Unfortunately, Ptolemy did not use Eratosthenes' measurements in his great work but chose the circumference estimated by a later Greek, Poseidonius (186-135 BC). His measurement was 180,000 stades, or approximately three-quarters of the actual distance. The result was that many centuries later, while Columbus had factual evidence of the round world, the measurements he used suggested that it was smaller than it really is.

PROOF OF A ROUND WORLD

In the hundred years or so before the fourteenth-century rediscovery of the scientific work of Greek scholars, European explorers began venturing to other lands. While the travelers made few direct observations about the size and shape of the earth, their accounts of their travels stimulated trade, which in turn proved significant in renewing the quest for facts about the earth. The Venetian explorer, Marco Polo, who made his first journey to China in 1271, was perhaps the most important figure in the reawakening of trade. His route across the continent of Asia and back through the China Sea, the Indian Ocean, and the Persian Gulf opened up new trade routes (Figure 1-4).

Marco Polo presented to Europe a rich portrait of the Eastern world. Stories of spice islands and great, rich cities where people dressed in silks attracted the attention of kings and merchants. Trade developed with the East, and the Italian merchant states had a monopoly on the caravans carrying spices and textiles across the land route. To break this monopoly, countries like Spain and Portugal looked for sea routes to the

Figure 1-4. Marco Polo's route from Venice to China and back. (From ALL POSSIBLE WORLDS by P.E. James, copyright (c) 1972 by The Bobbs-Merrill Company, Inc.)

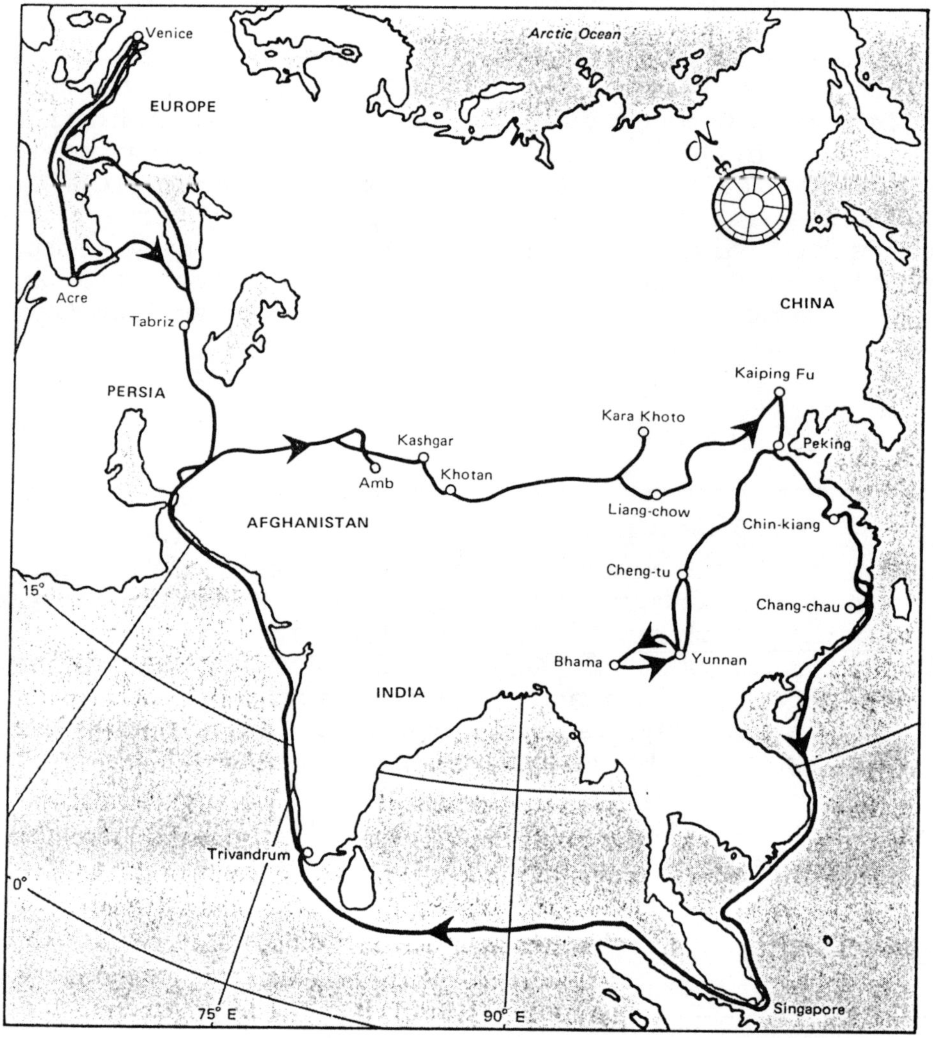

fabled East. By stages, at first through occupying islands in the Atlantic and then by pushing southward along the African coastline, an eastern sea route was eventually established. The improvements in navigation and ship design that accompanied these explorations set the scene for the search for a route west rather than east.

The story of Columbus's single-mindedness in searching for financial backing for such a venture is well known. Less well know is the theoretical background that he drew upon. One of the best sources of ideas available to Columbus was in Ymago Mundi (Image of the World) written by Pierre d'Ailly and published between 1480 and 1483. A copy of this text exists with Columbus's extensive marginal comments. So heavily annotated is the manuscript that the historian, Samuel Eliot Morison, suggests that it "seems to have been Christopher's bedside book for a number of years" [4:120]. A few examples illustrate how Columbus used the text. At the beginning of the book d'Ailly writes: "The world has a spherical or round figure with a variety of many parts." In the margins, Columbus's extensive notes include: "A sphere is called a space that is a rotundity or a body round on every side...the sphere has neither beginning nor end. And thus it is not easy to understand where what is round, as a circle, begins and ends" [2:361]. It is evident that Columbus was convinced that the earth was round.

Another marginal comment that played an extremely important role in Columbus's trips was his note on the size of the sphere. Amplifying d'Ailly's description, Columbus wrote: "Aristotle. Between the end of Spain and the beginning of India there is a small sea navigable in a few days." With such evidence it is little wonder that Columbus did not anticipate finding a new continent. Thus it was that in 1492 Columbus set sail with three ships to reach Marco Polo's Cathay and the island of Cipangu (Japan).

At this point the under-estimation of the earth's circumference became apparent. Columbus believed that Cipangu was some 5,600 kilometers (3,000 nautical miles) west of the Canary Islands but at about the same latitude. This is less than one-third the true distance. But he modified these figures himself and reduced the estimate by one-sixth, which meant that he located Cipangu in the longitude of the West Indies and Cathay on the coast of Mexico.

On his first voyage, after sailing for four weeks beyond the Canaries, he suspected that he had missed Cipangu and headed southwest for the mainland of Cathay. When he finally landed on Watling Island in the Bahamas, he had already passed the point where Japan was supposed to be by a week's sailing. In Cuba he actually sent a party to seek out the Great Khan. Even on his last voyage (1502-1504) Columbus still thought he was in Asia and identified the Central American coast as "Catigara" (Indochina) and believed himself to be but nineteen-days journey from the River Ganges in India. In fact, in his correspondence with the Spanish Royal Court, in which he begged forgiveness for failing to discover the riches he had promised, Columbus concluded the dispatch: "Written in India on the Island of Jamaica, July the 7th, 1503" [3].

In 1507, one year after the death of Columbus, the German cartographer, Martin Waldseemuller, used information from the Florentine explorer, Amerigo (Americus) Vespucci, to name the new world "America after the sagacious discoverer, Americus." Other explorers went to the New World, and its geography began to take shape. The acquisition of geographic information, combined with Magellan's trip around the world just a few years later, proved both that the earth was round and that its circumference had been greatly under-estimated.

HOW ROUND IS THE EARTH?

Confirmation of a round earth by explorers seemed, in itself, to close the issue of the earth's shape. In fact, it represented just the beginning of an entire new set of problems. The first problem was the difficulty mariners had locating themselves on the round earth. It was easy enough for mathematicians to construct a grid of latitude and longitude on the sphere, but how could location on the grid by established by a navigator aboard a rolling, pitching ship?

Eventually the solution to the problem of location on the round earth was found in a method of determining longitude by comparing local time with the time at a standard meridian. But in the sixteenth century the problem was so acute that most seafaring nations offered cash prizes to the person who could provide an answer. In 1598 Philip III of Spain offered pensions and gratuities to the value of 9,000 ducats for a solution to the longitude problems: Portugal, Venice, and Holland followed with similar rewards. No practical method resulted until 1732, when the English clockmaker, John Harrison, produced a remarkably accurate chronometer. Even then he had to wait until 1771 before gaining the prize of £20,000 that had been offered [1].

The quest for determining longitude provided some important observations concerning the shape of the earth, many of them resulting from work completed under the auspices of the French Academie des Sciences. As part of the Academie's research, scientist Jean Richer was sent to the island of Cayenne, off the coast of former French Guiana in South America, to observe the planet Mars. Part of his equipment was a highly accurate pendulum timepiece that had been carefully adjusted before he sailed. Richer found that in Cayenne the clock did not keep accurate time. In order to adjust it, he was forced to shorten the pendulum. When Richer reported this finding, it created much discussion but the explanation for the needed adjustment had to await the Publication of Newton's Principia in 1687. Newton suggested that the variation of pendulum swing in locations near the equator could be caused by a lessening of the pull of gravity resulting from a bulging of the earth in that region. He suggested that instead of a perfectly round sphere, the earth was an oblate ellipsoid, a geometric form that is flattened at the polar areas. The idea of the orange-shaped earth was introduced.

Not all scientists were convinced by Newton's explanation, and further survey work by Jean Picard and later by Jean and Jacques Cassini, all of the Academie, gave rise to the opinion that rather than being flattened, the polar areas were actually stretched, and the earth was a prolate rather than an oblate ellipsoid (Figure 1-5). To settle the argument the Academie sponsored expeditions to Lapland near the North Pole and to South America so that precise measurements of the length of 1° of latitude in different parts of the earth could by compared. The results substantiated Newton's oblate shape. Thereafter, many countries established survey systems, and accurate measurements of the earth were gathered in many places. One of the survey lines, completed in the United States in 1768 by Mason and Dixon, achieved another kind of fame in the Civil War.

Figure 1-5. Controversy over earth shape led to two depictions: the oblate and prolate ellipsoids. All of the angles shown are equal. (From R.K. Burkard, Geodesy for the Layman, 1964, p.5. Geophysical and Space Sciences Branch, Chart Research Division, Aeronautical Chart and Information Center, St. Louis, MO.)

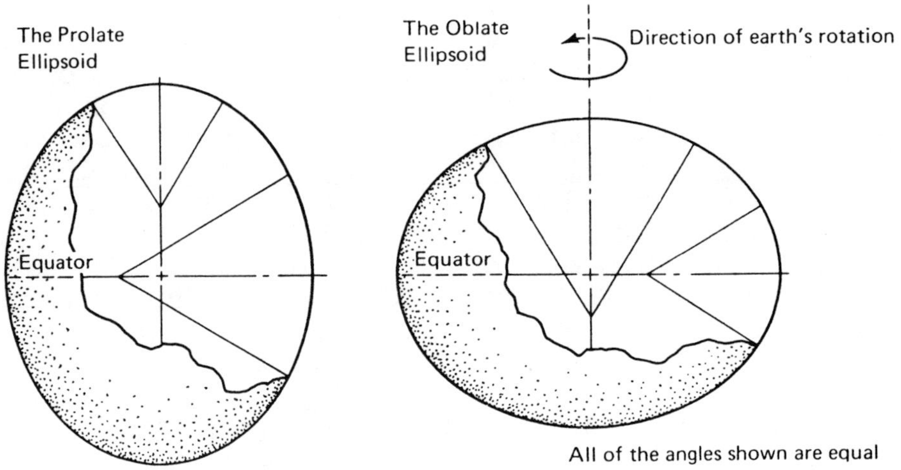

NOW A PEAR-SHAPED EARTH

With the oblate ellipsoid accepted as the earth's shape the next task for earth scientists was to measure the equatorial and polar dimensions so that the exact amount of polar flattening could be determined. By looking at Figure 1-6 we see that the amount of flattening is calculated from the difference in length between one-half of the equatorial diameter (the semimajor axis "a") and one-half of the polar diameter (the semiminor axis "b"). Precise measurements of the axes were required to establish information for a reference ellipsoid: the idealized geometric shape that best approximates the true form of the earth. The dimensions used in the ellipsoid form the basis of the exact lengths of degrees of latitude and longitude required to locate a point on the globe. The determination of data is no easy task, and, as a result, a number of reference ellipsoids have been calculated. The table in Figure 1-6 gives the dimensions of several ellipsoids. As a basis for the mapping of the United States, the measurements derived by A.R. Clarke in 1866 are still used.

In calculating the dimensions of the ellipsoid, land-based surveys, astronomical data and gravity measurements were used. In recent years a new approach has been made available through space programs. Earth satellites, placed in orbit so that the earth's center of gravity is one focus in an elliptical orbit, have been found to shift westwards (Figure 1-7). It seems that the bulge at the equator exerts a twisting force on the orbiting satellites. Observations based upon such satellite motions have been interpreted to suggest that the earth is actually oval-shaped at the equator. Another theory suggests that the orbital variation of satellites points to additional earth flattening in the south-polar area accompanied by an equivalent bulge in the north, while at the same time middle latitudes in the Northern Hemisphere bulge less than those in equivalent locations in the Southern Hemisphere. To visualize such an earth shape one might compare it to a fat pear (Figure 1-8).

Figure 1-6. **a.** The figure showing the parameters used to determine the earth ellipsoid. **b.** A table of the dimensions of selected earth ellipsoids. (From R.K. Burkard, Geodesy for the Layman, 1964, p.8. Geophysical and Space Sciences Branch, Chart Research Division, Aeronautical Chart and Information Center, St. Louis, Missouri.)

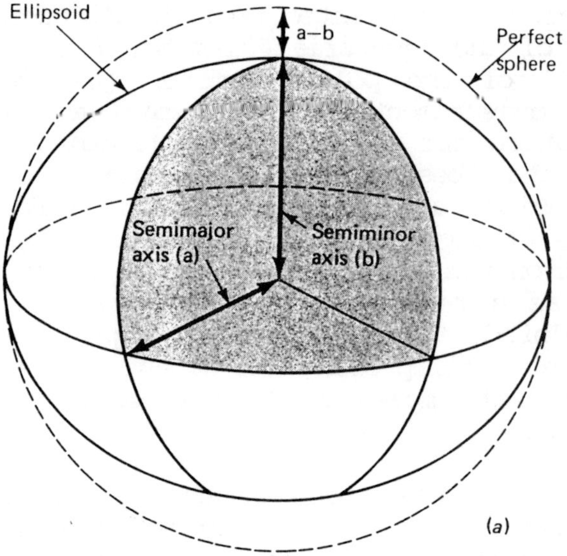

Dimensions of Selected Ellipsoids

Ellipsoid	Semimajor Axis[1] *a*	Semiminor Axis[1] *b*	Flattening[2] *f*	Fractional equivalent
Everest, 1830	6,377,276	6,356,075	0.003,324	1/300.8
Bessel, 1841	6,377,397	6,356,079	0.003,343	1/299.15
Clarke, 1866	6,378,206	6,356,584	0.003,390	1/294.98
Clarke, 1880	6,378,301	6,356,584	0.003,408	1/293.47
International (Hayford, 1909)	6,378,388	6,356,912	0.003,367	1/297.00
Astrogeodetic, or Mercury Datum (Fischer, 1960)	6,378,160	6,356,778	0.003,352	1/298.3
Geodetic Reference System (IUGG, 1967)	6,378,160	6,356,775	0.003,353	1/298.25

[1] *Measurements in international meters.*
[2] *Flattening (f) is given by* $f = \frac{a - b}{a}$

(b)

Figure 1-7. Because of the equatorial bulge, an orbiting earth satellite moving eastward will shift toward the west.

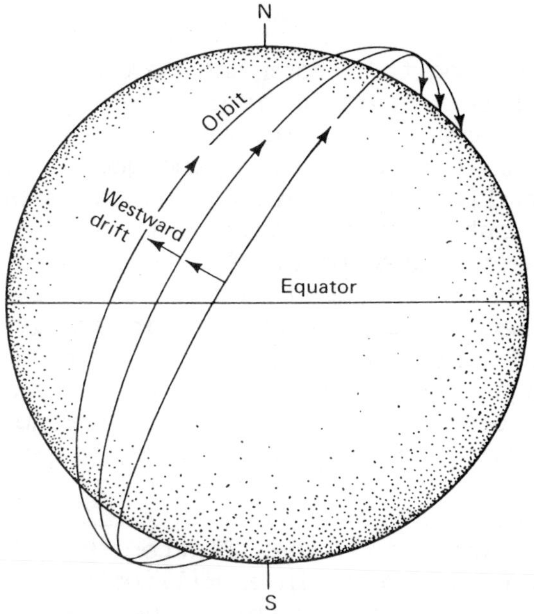

Figure 1-8. This depiction of the pear-shaped earth is based upon the earth's sea-level surface. Variations from the ellipsoid are greatly exaggerated.

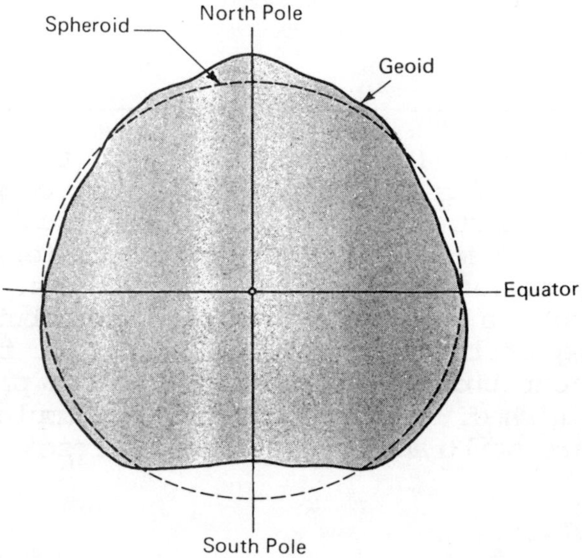

The study of the earth's form has become so specialized that it is now treated as a distinct area called geodesy. The modern geodesist looks beyond the elliptical shape of the earth and is more concerned with the form called the geoid. The shape of the geoid can be visualized by imagining thousands of sea-level canals that are cut across the continents and are allowed to fill with seawater. The level that the water attains (after eliminating fluctuations caused by tides and waves) defines the geoid surface. Because of differences in gravity over the earth and because of the effects of the earth's rotation, the water level, and hence the geoid surface, is not a smooth sphere. As we can see in Figure 1-8, the surface is irregular. Therefore, its surface differs from the geometric ellipsoid. Satellites are also being used to help define this complex geoid. In fact, satellite geodesy now makes it possible to determine the radius of the earth and the flattening of the equator's ellipsoid to about one part in a million and to determine the position of ground stations within nine meters (30 feet) with reference to the earth's center of gravity [6:436]. Particularly promising in studying the form of the earth is the laser geodynamic satellite now in use.

CONCLUSION

It is a long step from initial proof of the earth's roundness and the size of its circumference to the complexities of the geoid. The evolution has involved practical and theoretical contributions from people of quite different backgrounds. Astronomers, sailors, geologists, watchmakers, and geographers, to name but a few, have all contributed toward our understanding of the earth's shape. Without their studies and discoveries our understanding of many earth processes would be greatly diminished. The picture is not complete, however, and geodesists are still actively engaged in research.

SUGGESTIONS FOR FURTHER WORK

1. Suppose that you met a person who believed the earth was flat. How would you convince him/her that the earth is a sphere?

2. (a) In relation to everyday life how important is it to visualize the earth as a sphere? Is the spherical shape necessary to account for many of your observations of the environment in which you live? (b) Why should geodesists be concerned with the exact definition of earth shape? Do their findings relate in any way to human activities?

SELECTED SOURCES FOR FURTHER READING

Cotter, C.H. The Astronomical and Mathematical Foundations of Geography, New York: American Elsevier Publishing Co., 1966.

Hayward, O.T., et al. Dimensions of a Planet, (Council on Education in the Geological Sciences of the American Geological Institute: Programs Publication No. 7). New York: McGraw-Hill Book Co., 1971.

Lapp, H. Dynamic Earth: An Introduction to Earth Sciences. New York: McGraw-Hill, 1973.

Strahler, A.N. The Earth Sciences, 2nd ed., New York: Harper & Row, 1971.

Thrower, Norman J.W. "The Discovery of the Longitude", in Navigation, Volume 5, Number 8 (1957-58), pp. 374-381.

REFERENCES

1. Brown, L.A. The Story of Maps, New York: Bonanza Books, Crown Publishers, Inc., 1949.
2. Grants, E., ed. A Source Book in Medieval Science, Cambridge: Harvard University Press, 1974.
3. Leithauser, J.G. Worlds Beyond the Horizon, translated by H. Merrick, New York: Knopf, 1955.

4. Morison, S.E. Admiral of the Ocean Sea, Boston: Little, Brown and Company, 1942.
5. Tuan, Yi-Fu. Topophilia, Englewood Cliffs, N.J.: Prentice-Hall, 1974.
6. Veis, George. "Geodesy", in Encyclopedia Americana, International Edition, Volume 12, Danbury, Connecticut: Grolier Inc., 1986.

Chapter 2

Map Projections: Misconceptions and Propaganda

The Context

The study of spatial aspects of world physical, economic, and political relationships and their distribution goes hand in hand with the use of maps. Maps are two-dimensional representations of selected features of the earth's surface at reduced scales and, thus, have significance in geography because they bring a generalized reality of the phenomena and forms of the surface of the earth to a scale of human comprehension. Therefore, maps are a principal tool in geographic analysis.

Most people have, at one time or another, looked at a map of the world. Unfortunately, most maps they may have consulted do not show the true character of the earth because it is impossible to transfer accurately the features of a spherical surface, the earth, to a flat sheet of paper, the map. In some way the map will distort the true picture of the world. How it distorts the earth's surface depends upon the map projection used.

It is not surprising, therefore, that numerous representations of the earth's surface have been devised. To locate points on the earth's surface, the mapmaker employs a method which uses meridians and parallels to divide the earth's sphere into segments. This grid system is then transferred to the flat surface of a map where the spatial characteristics of the earth's surface, such as area, shape, direction, and distance are duplicated, at a particular scale. Area is the only one of these which can be accurately represented without qualification on the entire map. Depending on the particular properties that are most needed for the map's use, the map maker will choose a particular projection; however, while he can represent some of the properties singly or in limited combination, all of them can never be represented. The map maker thus chooses the projection which best maintains the map properties for his specific purposes. It is axiomatic, therefore, that all maps lie in some manner and that the map user needs to be aware.

The study of the construction of map projections of the earth requires a basic understanding of mathematics and is too technical for most tastes. For this reason, and because few persons ever need to make a world projection for themselves, such studies often appear only in book appendices. Yet the lack of understanding of the basic attributes of a map projection can lead to misconceptions about the earth. In this chapter we will examine one of the best known and most widely used of all map projections. The study provides a warning concerning the care that is necessary in interpreting facts from maps.

One of my lasting memories of the English schoolrooms, where I sat as a student, was "The Map." This was not any old map. It was a huge wall map of the world, showing countries that belonged to the British Commonwealth and Empire. The most remarkable thing about "The Map" was its color, for the Commonwealth countries were always shaded an intense pink. To me, it seemed as if the whole world was pink. It was not until many years later that it occurred to me that perhaps the image conveyed by "The Map" was misleading. I did not realize that it was impossible to show our round earth correctly on a flat sheet of paper. To a young mind, any map of the world is a true picture of the world. The picture that "The Map" gave was certainly not the true one.

DIFFERENT MAP PICTURES OF THE WORLD

"The Map" was drawn on what is known as the Mercator projection (Figure 2-1). Like any projection, it represents a method by which the latitude-longitude coordinate system of the round earth is transferred to a flat surface. It takes its name from the Flemish map maker, Gerhardus Mercator, who introduced it in 1569.

Study of the Mercator projection shown in Figure 2-2 indicates that the equator is a straight line that stretches entirely across the map. Latitude 60°N is also a straight line, and it is the same length as the equator. If you look at a globe, you see that the distance along the 60°N parallel is much shorter than the distance along the equator. In fact, it is only half as long. Consider now the north-polar area. If the pole were shown (and it is not), it would be represented by a line the same length as the equator! A single point, the North Pole, would be shown as a line some 25,000 miles long!

It is clear that on the Mercator projection the further from the equator, the more the world is stretched. Obviously, a map drawn on such a projection cannot show the true area of any country. Areas nearer the poles are enormously exaggerated in

Figure 2-1. The world shown on the Mercator Projection.

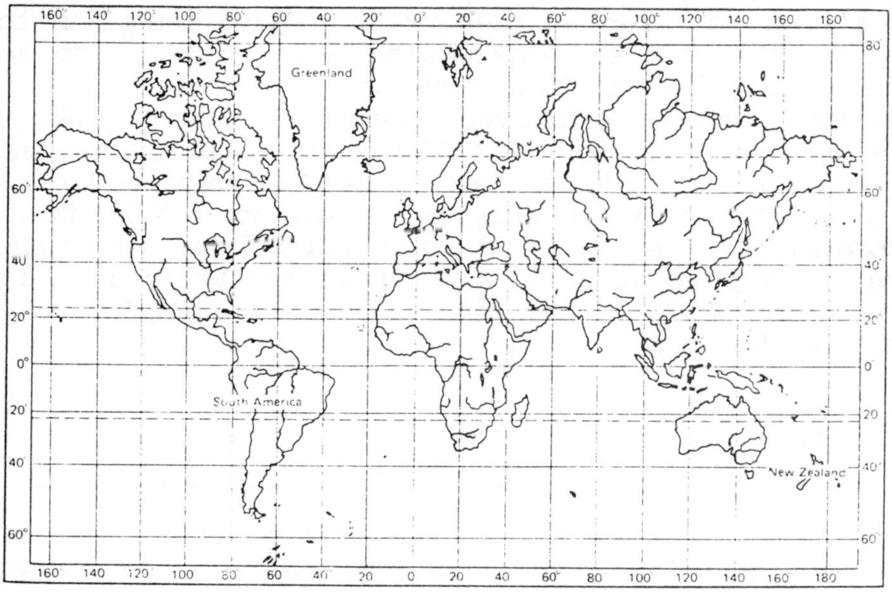

Figure 2-2. The world shown on an equal-area map, Goode's Interrupted Homolosine projection. (Goode's Map Base, used with permission of the University of Chicago.)

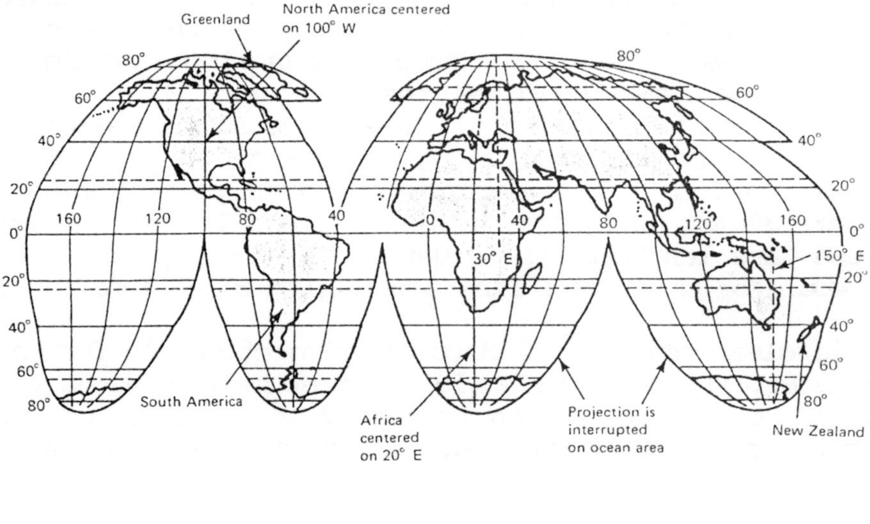

size. Compare, for example, the size of Greenland in Figure 2-1, with its size in Figure 2-2. Figure 2-2 represents a map projection whose main property is the preservation of equal area: the correct size of land masses is shown in relation to one another. Still, it is not a perfect map, for to show equal area the shape of the land is distorted. Nonetheless, the equal-area map shows quite a change in the size of polar land areas on the Mercator projection.

Since the Mercator projection produces such exaggeration, why was it devised in the first place? The answer lies in Mercator's explanation that accompanied the publication of his projection. "If you wish to sail from one point to another, here is a chart, and a straight line on it, and if you follow this line carefully you will certainly arrive at your destination" [1]. In effect, Mercator did produce a navigational chart of the world, for any straight line drawn on the map provides a rhumb line, or line of constant compass bearing. Mercator went on to say: "But the length of the line may not get you there as soon as you expected but you will certainly get there." He was pointing out that while a straight line on his projection does give the true direction, it certainly does not indicate the shortest distance, or the great circle route, between any two points on the earth. To plot the shortest route, navigators need to determine their routes from a projection that can provide true distance and then derive the compass bearings by plotting that route on the Mercator projection.

That the Mercator map became so widely used in teaching probably relates to its aesthetic appeal. It shows the world as a single continuous panel without all the curves and interruption of many other projections. What is more, the enlargement of areas toward the poles does make it easy to see small European countries like the British Isles, which on other projections, drawn at the same scale, would be barely visible.

A second feature of the "The Map" was that it showed the British Isles at the center of the world, with commonwealth countries like New Zealand appearing at the very lower edge. Such an arrangement of the land masses results from the fact that the Mercator projection is a simple, rectangular panel that is most frequently centered on the Greenwich meridian. The equal-area map in Figure 2-2 is not centered on this meridian, and the interruptions and curved lines of the projection give a different impression of relative world location. There is no obvious beginning and end of the map. Each continent, at least in visual terms, assumes equal importance and it is less easy to impose a focal point. The widespread use of other perspectives of the earth has helped to overcome the idea of the centrality of any single location. The fact that any place can appear as the center of the world is evident from the equidistant projections showing flying distances from one location to others. Such projections can be seen in many international airports. The center of the world could just as well be Karachi, Pakistan as London, England. Figure 2-3 is an

Figure 2-3. An equidistant projection of the world, centered on London. (From A.N. Strahler, Physical Geography, 4th ed. (1975), p. 27. Reproduced by permission of John Wiley & Sons, Inc.)

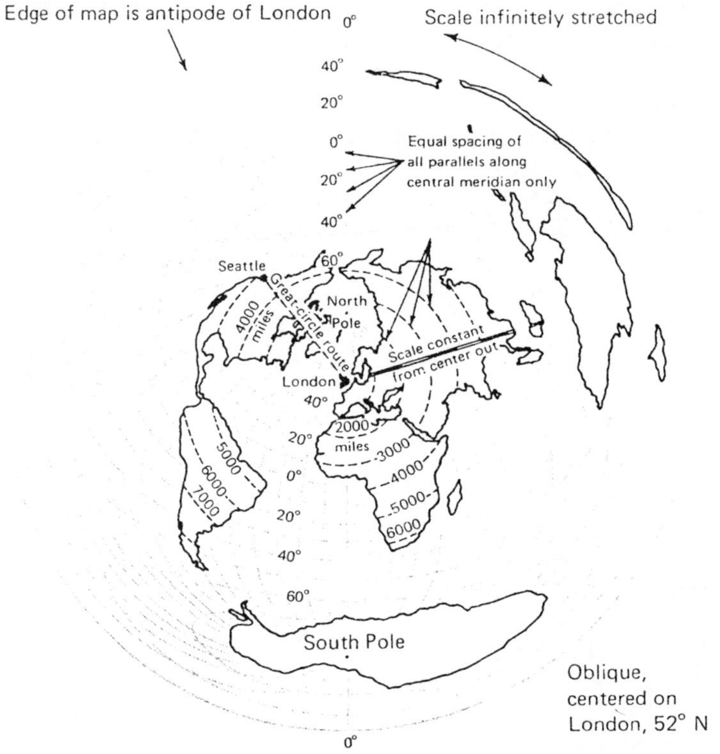

equidistant projection centered on London. Using the scale provided, the reader can, with a single measurement, determine the distance from London to any other place in the world. The English schoolroom is not the only place where the Mercator view of the world has been used. In fact, it has been suggested that such a world view was held in high government levels prior to World War II. A.K. Henrickson suggests some ways in which the Mercator projection influenced the thinking of American military strategists. Although diplomatic talks with the Japanese had degenerated and the United States anticipated military action by the Japanese any day, Americans were totally unprepared for the surprise attack on Pearl Harbor, Hawaii in December, 1941. The Japanese had crossed 3,500 miles of open sea without being detected.

Conceptually, the distance across the Pacific was even greater than this. By cartographic convention, most world maps were centered on the Greenwich meridian. This entailed the division of the Pacific Ocean into two parts, approximately along the International Date Line just west of Alaska, which put Hawaii on the left side of the map, in the extreme "West," and placed Japan at the extreme right, in the "Far East." For the Japanese to attack the United States directly, they would need figuratively to traverse the entire width of the map. On such world maps the distance between the Japanese and Hawaiian Islands was not directly apparent. It was equally difficult to relate them to one another directionally. This visual discontinuity surely contributed to the conceptual gulf that separated the United States and Japan [2].

As the war progressed, such a perception of the world could not persist, and a number of cartographers made important contributions to changing the widely held Mercator view of the world. Much of the innovative cartography that was produced, stems from the work of Richard Harrison, a technical adviser to the geographers of the State Department. His main idea was to portray the earth from as many different viewpoints (by using different map projections) as possible so that viewers would reconsider their perceptions of world relations. To attain a large audience, his maps were reproduced in magazines like Life and Fortune. The maps produced were frequently shown on an equidistant projection like that in Figure 2-3. While such maps set the scene for the "air-age" view of the earth, their main contribution has been to help change the Mercator image to a view of the world that more accurately reflects true geographic relationships.

MAPS FOR PROPAGANDA

Just as use of the Mercator projection contributed toward a rather biased view of world relationships, so it is possible to produce maps to literally make people see what you want them to see. Suppose you were a Russian interested in propaganda and wanted to show how much larger the Soviet Union is than the United States. The simplest possible method is to use the Mercator projection. The high-latitude location of the Soviet Union causes its size to be exaggerated many times (compare Figure 2-2 and 2-3). This simple method can be greatly enhanced by a few other modifications, like selective shading and deciding how much of the map to show in an illustration.

In his account, Maps and the Perception of Space, Roger Prestwich shows exactly how, by using bold shading, an area can be made to stand out on a map and therefore be misrepresented in its relationship to other areas [4]. Figure 2-4 illustrates how this shading might be done to exaggerate the size of communist countries. It uses the Mercator projection to again maximize the size of the Soviet Union. But large parts of the noncommunist countries are omitted in the map. If Greenland, Alaska, the southern tip of South America, and Antarctica were left in, the exaggeration in the projection would not be effective. The object of the map, enclosed in the inner box, is to show the threat of communism to the world. The viewer's eye must be drawn to the extent of the communist countries. These are shaded in bold diagonal lines. By contrast, the rest of the world is an innocent-looking light shade. Notice, too, how labels are added to the map. "World Communism Today" is imprinted over a large block of the free world. "Europe" merges into the shaded zone so that the actual extent of

Figure 2-4. The Mercator projection is used to show the threat of Communism to the Free World. (From D. Lanegran and R. Palm, eds., An Invitation to Geography. Reproduced by permission of McGraw-Hill Book Company.)

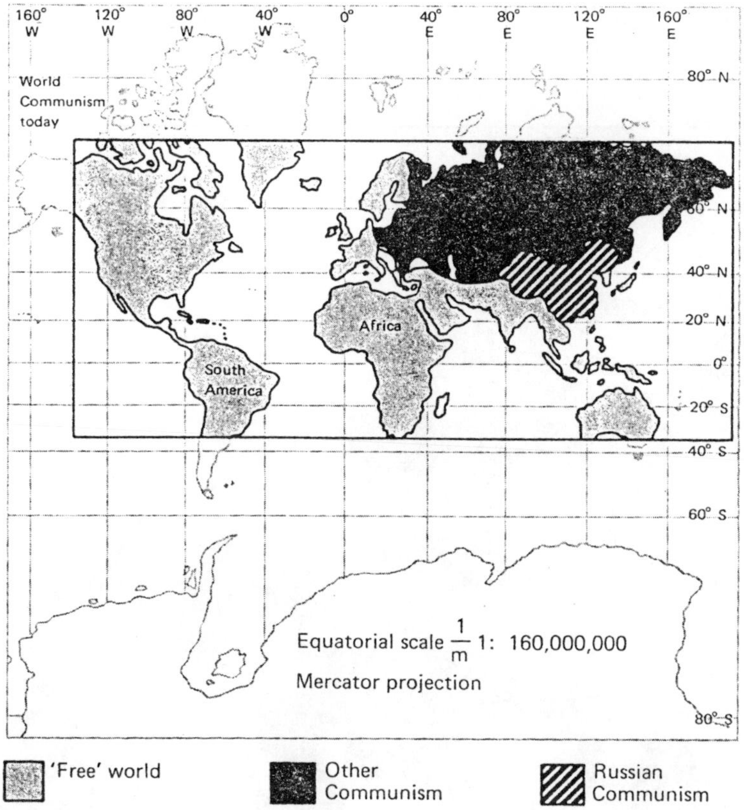

Eastern European communism is hidden. The threat is clearly seen.

Suppose now that a view from the Russian side is needed. A different projection might be employed so that the American threat of encroachment is shown (Figure 2-5). This time the heavy, eye-catching shading is put on the American areas. To emphasize the aggressive encirclement, American military bases

Figure 2-5. The American threat to Russia is depicted, in this map, drawn on a polar projection. (From D. Lanegran and R. Palm, eds., An Invitation to Geography. Reproduced by permission of McGraw-Hill Book Company).

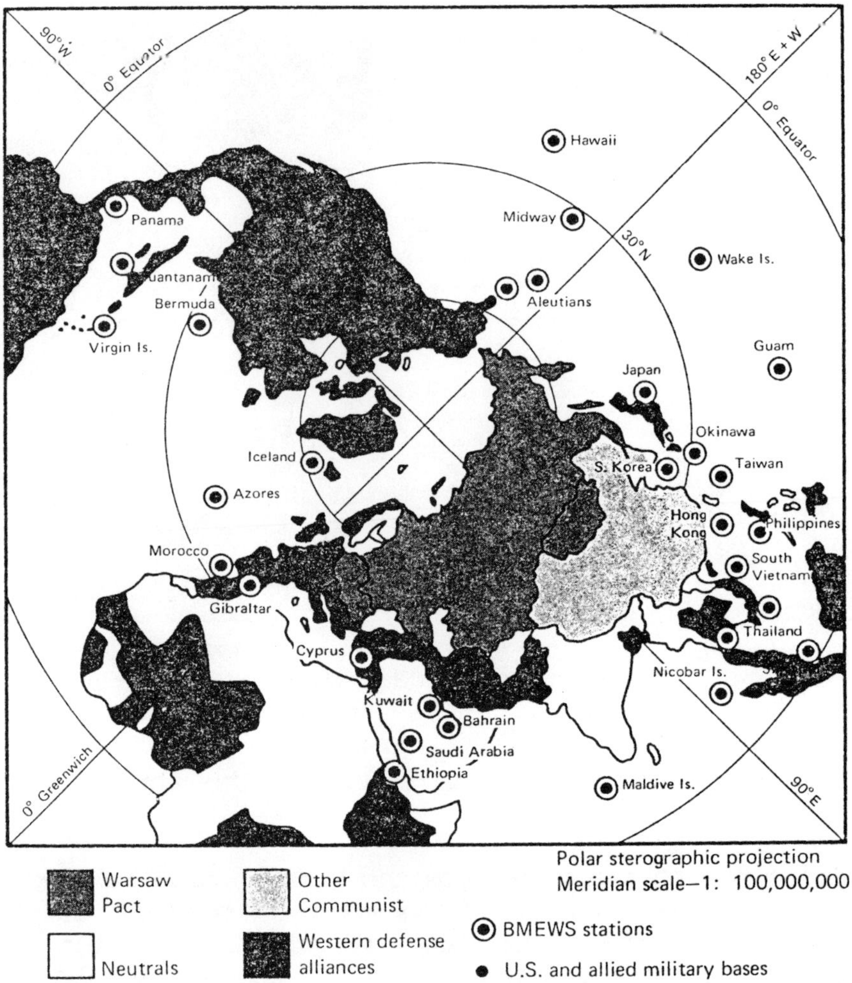

and missile sites are shown. Note how large the symbols are. The Ballistic Missile Early Warning System (BMEWS) stations are shown by symbols much larger than many small countries. Cartographically, the encirclement is complete.

The use of map projections with appropriate shading for propaganda purposes has been widely exploited. Probably, the most active exponents of the art were the German cartographers of World War II. A glance at some of their maps, such as those reproduced in an article by Louis Quam [5], shows just how expert they were in providing visual expression of political dogma. One item stressed by German propagandists was Hitler's concept of Lebensraum, living space for the German people. This concept arose, in part because the boundaries imposed after World War I incorporated groups of German-speaking people inside other countries. To achieve support for Lebensraum, the German people needed to be made "space conscious", and maps played an important role in this effort. As part of the propaganda program threat, maps were widely reproduced. A typical example, shown in Figure 2-6, was titled "Ein Kleinstaat Bedroht Deutschland," "A small state threatens Germany." Lines are shown radiating from Czechoslovakia to cover all of Germany. Printed over the lines are aircraft, indicating Germany's supposed vulnerability to air attack.

Another German map attempted to dissuade the United States from joining the anti-Nazi forces. Figure 2-7 is a world map, first published in a German propaganda booklet, showing "natural spheres of interest." The lines divide the world into three areas and imply that the United States is economically and strategically removed from the Euro-African and Asian realms and has nothing to gain from entering the war. Even a simplistic map, such as this one, provided material for those in favor of United States nonalignment in World War II. One final point of note, in this propaganda map, is the use of the Mercator projection. The Germans retained the common view of world relationships, for a new projection might not convey the message with such visual impact.

Figure 2-6. This German map, published prior to World War II, implies a tremendous air strike capacity of Czechoslovakia against Germany. (Adapted from L.O. Quam, "The Use of Maps in Propaganda," Journal of Geography, v. 42 (1943), p. 27. Reproduced by permission of the Journal of Geography.)

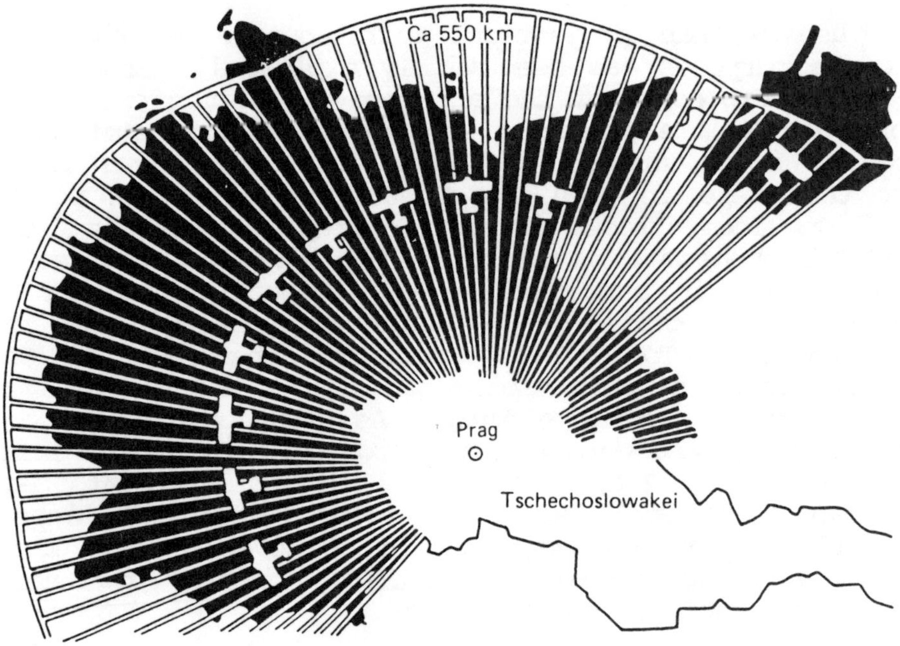

Ein Kleinstaat Bedroht Deutschland

Periodically, new projections capture the public's fancy for some purpose. Arno Peters announced, at a press conference in 1973, a "new" projection named in all modesty after himself (Figure 2-8). For the past decade he has promoted this projection, on the basis that it better represents all the peoples of the world, especially the three-quarters of the world's population who live in the Third World. He advocates the use of his map as the one correct representation of the sphere and claims that his projection can serve all purposes.

Figure 2-7. This map shows from left to right the natural spheres of interest, proposed by Nazi Germany, for the United States, Germany, and Japan. (Adapted from L.O. Quam, "The Use of Maps in Propaganda," Journal of Geography, v. 42 (1943), p. 30. Reproduced by permission of the Journal of Geography.)

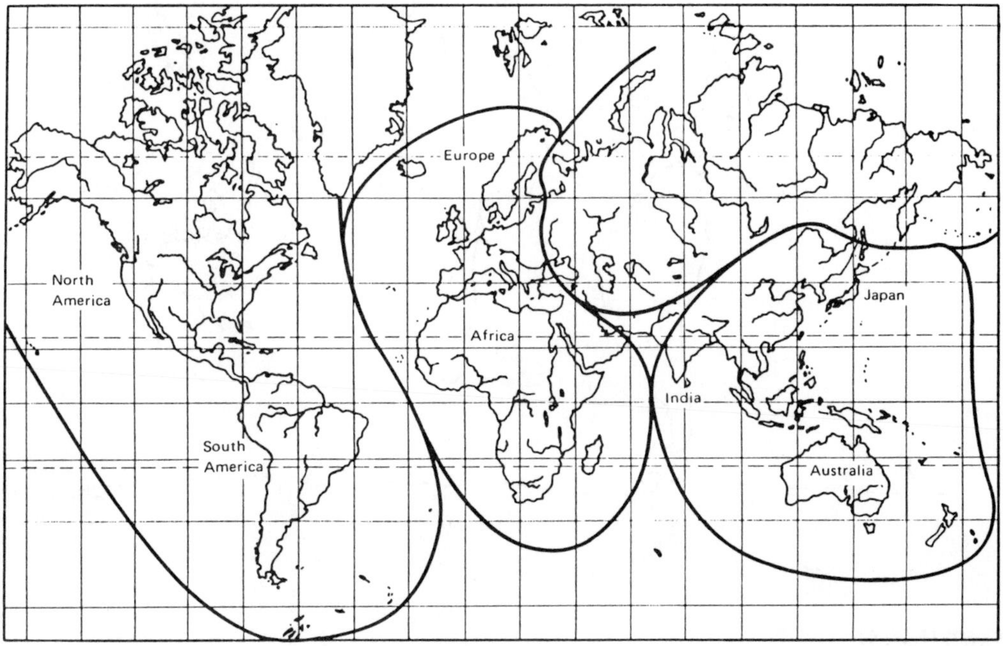

Because the World Council of Churches and other institutions have adopted his map, it has received some sanction. Unfortunately, many of the groups who need to use maps lack the technical background in cartography to evaluate Peters' propaganda. The projection is not new and was originally devised and presented by James Gall in 1855 (3). The map user needs to critically evaluate the claims and promotions of individuals such as Peters. Is any map best for all purposes? Since area and shape are mutually exclusive properties, what happens to shape in the Peters' projection? Does this map best represent the Third World or are there perhaps other better and more suitable alternatives?

Figure 2-8. Peters' projection (1973). with two standard parallels at 45 degrees. This equal area map shows great distortion of shape and distance relations in the tropics and at high latitudes.

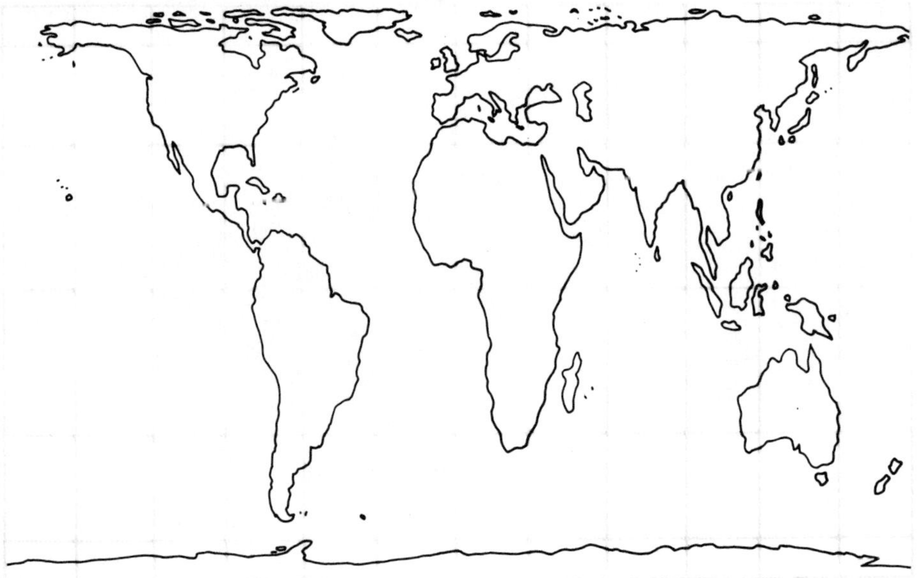

CONCLUSION

The examples we have given demonstrate the importance of the selection of map projections. The use of the Mercator projection was emphasized, and a few other projections were commented on briefly. The next time you look at a map, consider the projection used and the cartography employed. We have seen how projection and cartography are particularly important when maps are used in relation to geopolitics. There are numerous map views of the world, and each of these views preserves some truths but also has some drawbacks in its portrayal of the spatial patterns and relationships of the earth's surface. The user must be wary as the message being conveyed may or may not be a geographic reality.

SUGGESTIONS FOR FURTHER WORK

1. Although it is probably the best known, Mercator's projection is not the only one called by its inventor's name. Others include Gall's projection, Mollweide's projection, Goode's projection, and Miller's projection. For each of these projections, find out the basic properties and consider the purpose for which it might be best used.

2. Collect world maps published in newspapers and news magazines and examine how shading and printing are used to convey propaganda.

SELECTED SOURCES FOR FURTHER WORK

Campbell, J. Introductory Cartography, Englewood Cliffs, New Jersey: Prentice-Hall, 1984.

Monmonier, M. and G.A. Schnell. Map Appreciation, Englewood Cliffs, New Jersey: Prentice Hall, 1988.

Raisz, E. Principles of Cartography, New York: McGraw-Hill, 1962.

Snyder, J. "Map Projections - A Working Manual", Professional Paper 1395, Washington: U.S. Geological Survey, 1987.

REFERENCES

1. Greenhood, D., Mapping, Chicago: University of Chicago Press, 1965.

2. Henrickson, A.K., "The Map as an Idea: The Role of Cartographic Imagery During the Second World War," The American Cartographer, 2 (1975), pp. 19-53.
3. Porter, Phil and Phil Voxland, "Distortion in Maps: The Peters Projection and Other Devilments," Focus, pp. 22-30, Summer 1986.
4. Prestwich, R., "Maps and Perception of Space," in D.A. Lanegran and R. Palm, eds., An Invitation to Geography, New York: McGraw Hill, 1973.
5. Quam, L.O., "The Use of Maps in Propaganda," Journal of Geography, 42 (1943), pp. 21-32.

Chapter 3

Views of the Earth

Remote Sensing

The Context

Comparison of the technology available today with that of fifty years ago shows remarkable changes. The advances have affected all forms of the life-styles, and it is difficult, for example, to envision a home without a television set or a business without a computer. Of the many advances perhaps none is more spectacular than the journeys into space. The discoveries of the space age have added totally new perspectives to the world of science and our views of our planet.

Today, orbiting satellites routinely provide images and data about the earth that could not have been imagined a few decades ago. Further, without the availability of computer technology, the mountain of information acquired from instruments aboard satellites simply could not be handled. The questions taken up in this chapter concern the types of data that are being generated and how the environmental scientist and physical geographer have developed new techniques to analyze them. Remotely sensed data provide important inputs into studies of the physical and cultural patterns that exist on earth.

Remote sensing is a term with space-age connotations. Complex sensors orbit the earth and beam their acquired data to equally complex ground receivers. Yet, by stopping to consider the term in its broadest sense, we can see that it is really very simple. Remote sensing refers to the acquisition of information about an object from some distance away. When so defined, it is evident that all sighted persons have remote sensors. The optical instrument used in human remote sensing is the eye. It is a remarkable instrument. Equipped with an automatically adjustable pupil, its aperture varies with the level of illumination from 2mm to 8mm. The eye provides an automatic focusing mechanism from about 25cm to infinity and can distinguish objects at varying levels of brightness. Thus, this optical instrument allows a person to observe the world from afar; it is a remote sensing device in its truest sense.

Like the eye, the ordinary camera using panchromatic or color film is a remote sensing instrument. The photograph goes one step beyond the human eye, for it can freeze an image and, by selective use of filters, provide details not readily observed by the eye. Not long after the camera was invented its potential for recording the bird's-eye view of the earth's surface was recognized, and aerial photography was underway. Possibly the first aerial photograph was taken from a balloon by Gespard-Felix Tournachon in 1855 in France. By 1860, aerial photographs were taken over Boston. Since then, with the continual refinement of aircraft and camera equipment, aerial photography has become an integral part of the methodology used in interpreting the earth's surface. Flying heights now range from a few hundred meters to several hundred kilometers, involving, respectively, light aircraft and orbiting manned spacecraft.

One of the more significant instruments recently developed is the Large-Format Camera (LFC), first flown in 1984. This is a special purpose precision mapping camera which can be carried aboard the Space Shuttle. It can also be mounted in an unmanned spacecraft, with film to be changed by Shuttle astronauts, or mounted in an aircraft. The LFC has a 305mm focal length lens, and it produces photos 230 x 460mm in size with about three times the detail of conventional mapping cameras. In addition its advanced image motion compensation mechanism makes for high geometric fidelity, and stereoscopic coverage has been achieved as well [5:67-68].

ORBITAL PHOTOGRAPHY

The beautiful photographs of the earth taken from space are much more than pretty pictures. They represent another form of remote sensing that might be termed orbital photography. The significance of the photographs lies in the fact that a single frame covers areas of many thousands of square miles and permits the analysis of large features and structures of the earth's surface. The mile-by-mile mapping of surface faults by earthbound geologists, for example, has been shown to miss features apparent in orbital photographs. This was demonstrated in the northern mountains of Baja California and in southern California where photographs taken from Apollo IX showed a number of unmapped faults that provided new insights into the regional geology of the area.

Paul Lowran offers a number of interesting cases where orbital photography has helped to identify surface features and resolve old controversies [6:76-86]. One case concerns the discovery of the Palomas Volcanic Field in Chihuahua, Mexico. Gemini IV astronauts had obtained a cloud-free series of photographs extending from Baja California to central Texas, and a comparison of the photos with geologic maps for the state of Chihuahua showed a series of extinct volcanoes not previously identified. A search through earlier literature indicated that the volcanoes had never before been described, but--as a ground reconnaissance trip proved -- they did exist. The space photographs again showed how easy it was in earthbound surveys to miss obvious land features. The discovery was even more remarkable because the volcanoes were located in an easily accessible area, only two-hours drive into Mexico from the United States border.

REMOTE SENSING AND NONVISIBLE LIGHT

Both the human eye and panchromatic film have limitations in remote sensing. Both are able to sense only one small portion of the total range of radiant energy. This range is depicted by the electromagnetic spectrum: a continuum where various types of radiation are distinguished by the characteristics of the wave patterns by which they are transferred from their source. Thus X-rays and infrared radiation are both forms of electromagnetic energy, but they are differentiated from one another by the length of their waves. As shown in Figure 3-1, X-rays have a wavelength as short as 10^{-11} meters, while radio waves can be as long as 10^5 meters. Our eyes have receptors that can sense only a very small part of the spectrum, the part we define as the visible band. However, just as our eyes tell us about objects by using radiant energy in the visible band, so other kinds of sensors can use energy from other parts of the spectrum to bring us information we cannot obtain with our own eyes.

Figure 3-1 represents a portion of the electromagnetic spectrum, with the principal sub-divisions (bands) pictured in their relative positions, but not drawn to scale. Wavelengths are shown in metric units -- Angstroms (one ten-billionth of a meter) for the shortest waves and kilometers (thousands of meters) for the longest. Along the lower edge of the spectrum's representation, selected sensors are listed according to the wavelengths to which they are sensitive. Note that if "human eye" were added to Figure 3-1, it would appear in the visible part of the spectrum, along with "film in cameras".

Figure 3-1. A section of the electromagnetic spectrum showing examples of instruments used to monitor radiation at various wavelengths.

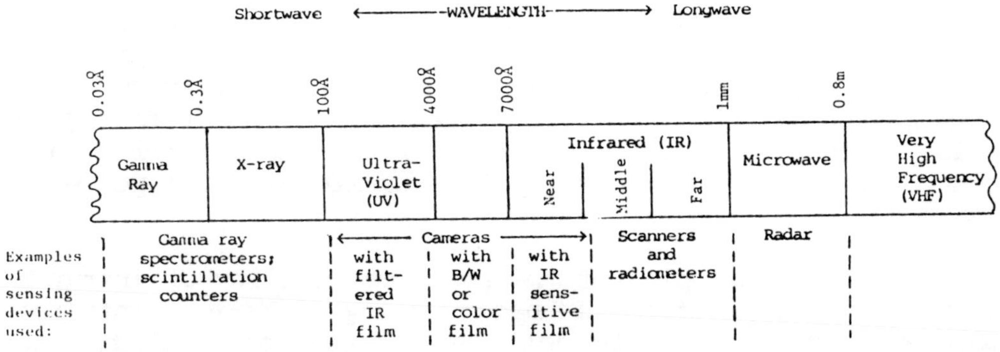

Black and white or color film in cameras responds to energy reflected from the various objects in the scene. For example, in an aerial photograph an equatorial rainforest usually appears dark (since little light is reflected) while an area of bare rock, usually more reflective, appears in much lighter tone. In other words, "rocks" and "foliage" differ in their response to light, so on the photograph we can readily distinguish between them.

If we substitute infrared film for panchromatic film, however, we find that it is sensitive to radiation in wavelengths too long to affect the B/W film. Thus, different sensors can record different kinds of information. Infrared film is sensitive to radiation in the "near IR" part of the spectrum (Figure 3-1). This film was developed as part of the military strategy. Ground installations could be camouflaged by draping an airplane, for example, with a cloth painted to match the color of the natural setting. From the air. then, it was impossible to visually distinguish the target. However, green-painted cloth does not look the same as vegetation at other wavelengths, particularly in the near infrared. At that wavelength the camouflage reflects energy much more strongly than healthy vegetation does, and the two can easily be distinguished.

The film involved was appropriately named "camouflage detection film". With a refinement of equipment, researchers later developed color infrared film. On photographs produced from this film, surface features are seen in a color that differs from that seen in nature. The film is thus referred to as false-color film [4]. If false-color film can differentiate military targets, it can clearly be adapted for more peaceful purposes.

There are many examples of the practical utility of infrared sensing. In a study of the Tehuacan Valley of Mexico, infrared photography was valuable in delineating features invisible to the eye and not shown on panchromatic film [2:520-527]. The Valley has been continuously occupied since 7000 BC, and archaeologists are interested in tracing the culture there over that long period. Infrared color photographs clearly delineated former areas of habitation, now under crops, because of their different images on infrared film. Similarly, old boundary features, whether of stones or different vegetation types, were sharply delineated by the infrared radiation reflected into the camera. This form of remote sensing holds great promise for the detection of ancient cultural features.

A study of citrus groves in Florida showed that it is possible to identify diseased trees using infrared film [8:59-63]. The identification depends upon the fact that diseased or dying foliage results in a decrease in infrared radiation (Figure 3-2). The infrared film detects the change in the leaves before it is visible to the human eye so that effective treatment can begin at early stages of the disease. Great strides in the detection of disease in many field crops, such as cotton and wheat, have also been made using infrared film.

Figure 3-2. This black and white reproduction of a color infrared film shows the early stages of decline of citrus trees infected by burrowing parasites (see boxed areas). (Courtesy of G.G. Norman, Florida Department of Agriculture.)

THERMOGRAPHY

Figure 3-1 associates infrared film with the "near" or "reflective" part of the IR band. Sensing in the far (i.e., the thermal) part of that band requires a different device, one which responds to slight differences in temperature. All objects at temperatures above absolute zero (-273°C or -459.6°F) radiate energy, but the type of energy they radiate depends on their temperature. The sun, with a radiating temperature on the order of 6,000°C, radiates energy in the form of light, while objects at lower temperatures, like the earth or the human body, radiate heat energy. This heat falls in the far infrared portion of the electromagnetic spectrum and is not sensed by our eyes or by infrared film. (However, IR film can record temperature differences when these are extreme, as in the case of a fire.) Detecting heat energy being radiated from objects of relatively low temperature requires other sensors, namely, either radiometers or scanners.

It is understandable that the United States Forestry Service has given much attention to a sensing program for the detection of forest fires. As part of Project Fire Scan, a test area comprising some 8,200 square miles (21,238 square kilometers) in western Montana and northeastern Idaho was used in 1970 to evaluate the efficiency of remote sensing techniques for fire detection [3:2253-2259]. The object was to detect fires not visible to tower observers before those fires erupted into forest disasters. Since the need was to find areas of very high temperature, IR film could be used. Figure 3-3 shows two photographs of the same forest area in Montana. Photograph a was taken by a conventional camera with standard film. Photograph b was taken by a conventional camera with infrared film; it shows eight small smudge pots in the lower left corner that are not visible on the standard film or to the human eye. Results of the study show that of the 406 fires that occurred while the aircraft was in service, 103 were detected by infrared photography. Of those not detected, 303 in all, 203 were already being fought prior to the Fire Scan Flight, leaving 100 that were missed by the aerial survey. Despite the substantial number missed, the test was considered a success, and with improved equipment such sensing will be utilized widely in years to come.

UN-MANNED SPACECRAFT

Among the spacecraft serving as platforms for remote sensing of the earth environment, some of the most useful have been the

Figure 3-3. Made with standard camera and film, Photograph (a) shows a forest area in Montana. Photography (b), made with infrared film, shows the same forest areas. In the lower left corner notice the eight small smudge pots (the heat source) not visible in the first photograph. (Courtesy of U.S. Forest Service.

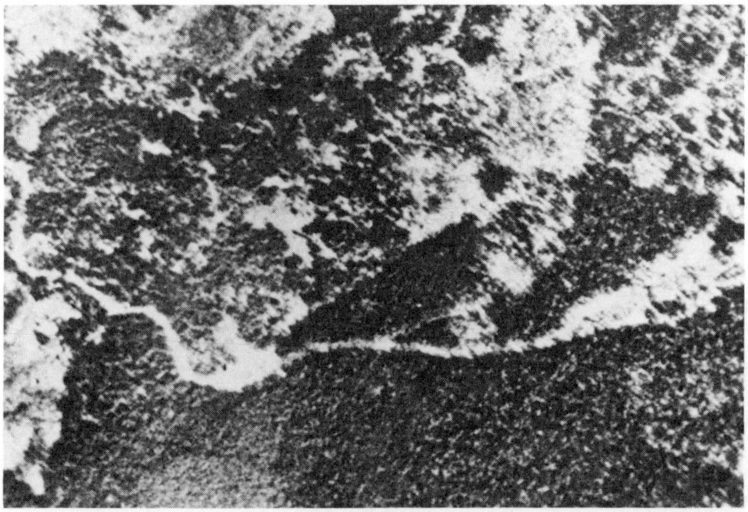

(a)

(b)

Geostationary Operational Environmental Satellites (GOES) operated by the U.S. National Oceanic and Atmospheric Administration (NOAA), since 1974. These make up part of a civilian global network of meteorological satellites. Two GOES orbit at about 36,000 km (22,300 miles) above the equator and synchronously with the earth's rotation. Thus, one satellite remains above the Atlantic and the other above the Pacific, high enough to view events in the atmosphere and the ocean in an entire hemisphere at one time. Twice each hour scanners aboard the satellites automatically generate one image in a thermal IR band and another in a visible band. These images are transmitted to earth receiving stations and relayed to local meteorologists who find them especially valuable in forecasting destructive storms. GOES scenes are familiar to millions of television viewers. Less well known is the fact that these satellites also collect raw information from river gages, buoys, ships, and other sensors for transmission to data processing centers [5:601-603;9].

Another satellite operated by the NOAA carries the agency name. Originally called Television and Infrared Observation Satellite (TIROS), NOAA was the first earth observation satellite. First launched in 1960 as an experimental weather satellite, the sun synchronous satellite has had a series of successors with a range of onboard sensors. Today the primary sensor onboard the NOAA satellites is the Advanced Very High Resolution Radiometer (AVHRR). This instrument collects data at 12-hour intervals at a spatial resolution of 1.1 kilometers in the visible, near infrared, and the thermal parts of the electromagnetic spectrum (Figure 3-1 and Table 3-1). To permit broader areal coverage for a given volume of data, an onboard computer can generalize data to a 4 kilometer resolution before data are transmitted to ground stations.

TABLE 3-1

Spectral Channel Characteristics of the AVHRR Sensor

Channel	Spectral Range	Spectral Region
1	0.55- 0.90 µm	Visible
2	0.72- 1.10 µm	Near Infrared
3	3.55- 3.93 µm	Thermal Infrared
4	10:50-11.50 µm	Thermal Infrared

One of the more recent applications of AVHRR data has been to monitor change in vegetation patterns over large geographic areas. Channels 2 and 1 can be ratioed using the equation below to form a greenness index (GI) of vegetation where:

$$GI = \frac{\text{Channel 2} - \text{Channel 1}}{\text{Channel 2} + \text{Channel 1}}$$

Such indices are particularly valuable for measuring the dynamics of major ecological zones. However, the satellite image's rather coarse resolution limits its utility to analyze specific sites [1].

The first Earth Resources Technology Satellite (ERTS-1), dedicated to earth surface analysis, was launched by the National Aeronautics and Space Administration (NASA) in 1972. Renamed Landsat in 1973, the fifth of these sun synchronous satellites, Landsat 5, is now in operation. The primary sensor on Landsats 1-3, and still included on Landsats 4 and 5 is the multispectral scanner (MSS) (Table 3-2). The MSS system includes an oscillating mirror scanning the terrain from east to west to produce a ground swath 185 kilometers wide. The pixel or ground resolution cell of data telemetered to earth from the satellite is approximately 79 x 56 meters, recording 128 different levels of brightness in each of the four bands. At

TABLE 3-2

Landsat Multispectral Scanner Channel Characteristics

Channel	Spectral Range	Spectral Region
1	0.5 - 0.6 μm	Green
2	0.6 - 0.7 μm	Red
3	0.7 - 0.8 μm	Near Infrared
4	0.8 - 1.1 μm	Near Infrared

this resolution one MSS scene contains approximately 31 million pixels of data. Various micro-computer-based digital image processing systems have been developed during the 1980's that allow for low cost and efficient processing of these large data sets.

Landsat 4, launched in 1982, initiated a new generation of the Landsat series. This satellite included a seven-band thematic mapper (TM), a sensor with spectral sensitivity and spatial resolution of 30 meters (about 100 feet)--better than those of the MSS (Table 3-3). Digital values of 256 different brightnesses can be recorded in each of the seven bands.

TABLE 3-3

Landsat Thematic Mapper Channel Characteristics

Channel	Spectral Range	Spectral Region
1	0.45 - 0.52 μm	Blue-Green
2	0.52 - 0.60 μm	Green
3	0.63 - 0.69 μm	Red
4	0.76 - 0.90 μm	Near Infrared
5	1.55 - 1.75 μm	Mid Infrared
6	10.40 - 12.50 μm	Thermal Infrared
7	2.08 - 2.35 μm	Mid Infrared

Landsat data can be used for monitoring a wide range of environmental processes. Vegetation analysis, land use change, geological structure, water resources, ocean surface characteristics, and environmental pollution are some of the major fields of environmental science utilizing Landsat data. Figure 3-4 is a multispectral Landsat scene of a portion of the Kansas High Plains, a region experiencing serious groundwater depletion. Landsat was used to monitor change in land used for various irrigated crops. The Landsat-derived data were merged with weather data and crop water consumptive use data to predict groundwater demand [7]. In another example, Figure 3-5 is a Landsat thematic mapper scene of Konza Research natural Area in the Kansas Flint Hills. Various combinations of bands provide useful biomass data allowing insights into the impact of burning and grazing strategies on one of the few remaining tallgrass prairie areas in North America.

Control of the Landsat program was transferred to the Earth Observation Satellite Company (EOSAT) in 1985 as part of an effort to reduce the federal government's direct role in the space program. Landsat data and imagery are available by contacting EOSAT, EROS Data Center, Sioux Falls, South Dakota 57198.

The success of Landsat has stimulated interest by other countries in similar programs. In 1986 the French launched SPOT (Système Pour l'Observation de la Terre). SPOT is operated by CNES (Central National d'Études Spatiales) in Toulouse, France. SPOT, a sun synchronous satellite, with successive overflights every 26 days, carries two identical sensing instruments. The two sensors, called HRV (High Resolution Visible) utilize a pushbroom scanning technology based on charged coupled devices (CCD's) that are capable of recording an entire line of data without mechanical movement [1] (Figure 3-6).

Figure 3-4. This computer classification of Landsat multispectral data provides information on the number of center pivot irrigation systems and associated crop types in northwest Kansas.

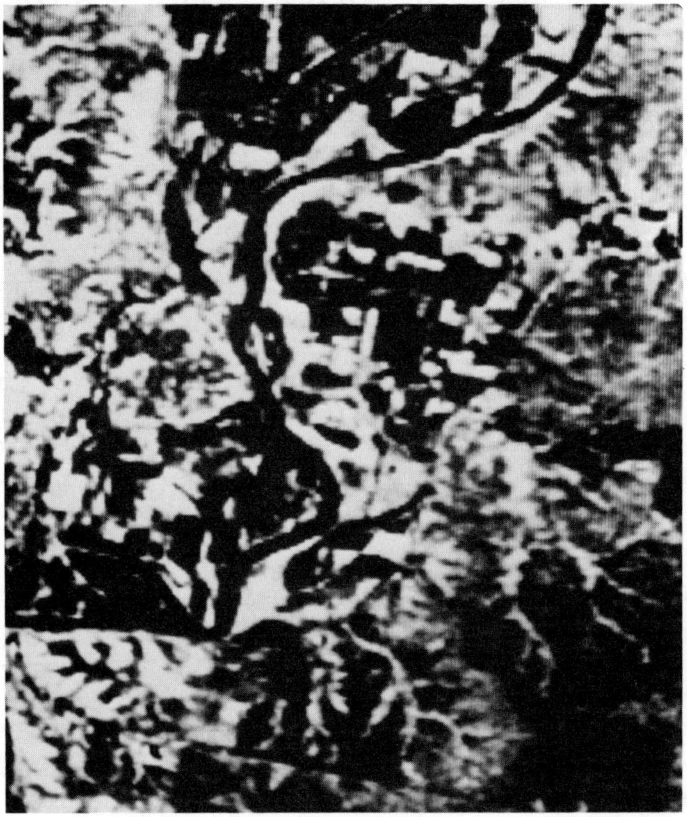

The HRV operates in two modes: panchromatic and multispectral. The panchromatic has one broad band extending from 0.51-0.73 micrometers. It images a 60-kilometer (36-mile) swath with a 10-meter (33-foot) spatial resolution. The multispectral mode also cuts a 60-kilometer swath across the earth's surface, sensing in three spectral regions: 0.50-0.59 micrometers (green); 0.61-0.68 micrometers (red); and 0.79-0.89 micrometers (near infrared). The multispectral mode records at a spatial resolution of 20 meters (63 feet). One interesting

Figure 3-5. Landsat data provided input for this image, showing variations in the greenness of vegetation as an index of biomass. Lighter tones correlate with high levels of biomass in the Konza Research Area in the Kansas Flint Hills.

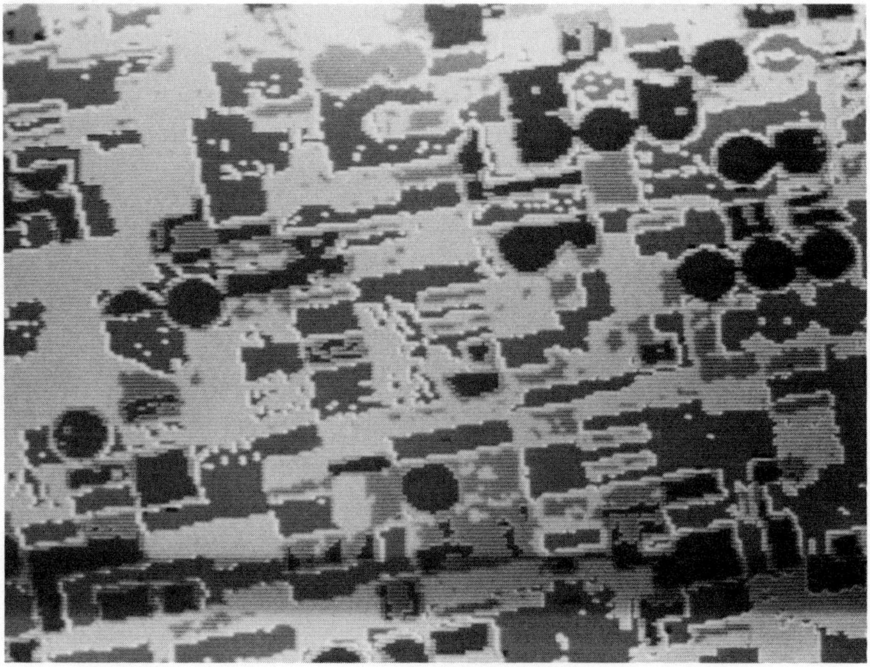

feature of the HRV modes is that they have off-nadir viewing capabilities. In other words, to gain more frequent data sets, it is possible to point the sensor back to an area of interest on earlier orbits.

Figure 3-6. Technicians working on the SPOT satellite, a space vehicle designed to monitor earth resources. (Courtesy of SPOT).

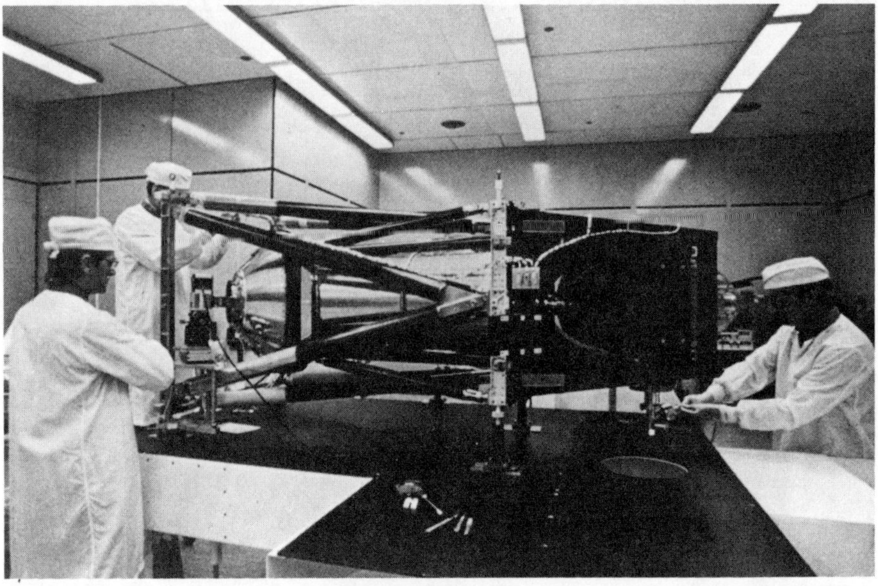

Applications of SPOT data to environmental assessment are similar to those of Landsat. Examples include structural mapping, hazards analysis, tectonic studies, land use mapping, crop forecasting, and pollution monitoring. SPOT data and or imagery can be acquired from: SPOT IMAGE CORPORATION; 1897 Preston White Drive; Reston, Virginia 22091-4326.

CONCLUSION

Remote sensing can provide opportunities not normally available to physical geographers. The ability to observe reflectance and emittance over a range of electromagnetic wavelength values opens the door to the study of subjects that would not otherwise be possible. The synoptic view of satellite images, for example, gives a broad scale perspective of patterns not discernible within the confines of the large-scale view from the ground. Images integrate shadow, soil, rock and vegetation information that can reveal subtle differences in terrain. Although remote sensing cannot replace direct ground measurement, it can serve as another source of information about our complex earth.

SUGGESTIONS FOR FURTHER WORK

1. This chapter includes some examples of the way thermal infrared data are applied to actual problems. However, those data have many other applications that were not discussed. Suggest some practical uses of thermal infrared imagery in the fields of energy resources, hydrology, oceanography, and plant geography.
2. The satellites described in this chapter are all classed as non-military. Should the information gathered while passing over one country be freely shared with people outside that country? Why? Explain your position.

SELECTED SOURCES FOR FURTHER READING

Campbell, J.. Introduction to Remote Sensing, New York: Guilford Press, 1987.

Curran, P. Principles of Remote Sensing, London: Longman, 1985.

Earth System Science Committee. Earth System Science: A Closer View, Washington: NASA, 1988.

Lillesand, T. and R. Kiefer. Remote Sensing and Image Interpretation, New York: John Wiley and Sons, (2nd Edition), 1987.

REFERENCES

1. Campbell, J.. Introduction to Remote Sensing, New York: The Guilford Press, 1987.
2. Gumerman, G.J. and J.A. Neely. "An Archaeological Survey of the Tehuacan Valley, Mexico: A Test of Color Infrared Photography", American Antiquity, Vol. 37, 1972.
3. Hirsch, S.N., et al. "The Bispectral Forest Fire Detection System", Proceedings of the 7th International Symposium on Remote Sensing of the Environment, Vol. 3, Ann Arbor: University of Michigan, 1971.
4. Holz, R.K., (ed.) The Surveillant Science: Remote Sensing of the Environment, Boston: Houghton Mifflin, 1973.
5. Lillesand, T. and R. Kiefer. Remote Sensing and Image Interpretation, New York: John Wiley and Sons, (2nd Edition), 1987.
6. Lowran, Jr., P.D.. "Space Photography--A Review," Photogrammetric Engineering, Vol. 31, 1965.
7. Nellis, M.D. Estimating Water Demand in Northwestern Kansas, Manhattan, Kansas: Kansas Water Resources Research Institute, 1987.
8. Norman, G.D. and N.L. Fritz. "Infrared Photography as our Indicator of Disease and Decline in Citrus Trees", Proceedings of the Florida State Horticultural Society, Vol. 78, 1965.
9. U.S. Department of Commerce, National Oceanic and Space Administration. GOES (Brochure), Washington, 1981.

Part II
The Atmosphere

Chapter 4

Drought in the Sahel

The Context

World population is increasing at an alarming rate. The rate of increase does vary, however, from country to country so that great contrasts are found between those areas where the population is beginning to stabilize and those areas where the number of people to feed, house, and clothe is outstripping the available resources.

In many parts of the world any deviation from the anticipated weather patterns creates dire hardships. In India, the timing of monsoon rains is a critical factor in determining the future of crops that feed millions of people. In northeast Brazil, failure of rains results in the massive emigration of farmers who cannot sustain life in a marginally moist region. Although the problems are felt so keenly at the individual level, their cause in part results from erratic patterns of weather associated with the general atmospheric circulation of the earth. It is a system that even today is not fully under our control.

In no area of the earth's surface is this pressure on resources more evident than in the 45 countries that make up sub-Saharan Africa. Here, population growth approaches three percent annually -- the world's highest rate. One-half to three-quarters of the people do not have a calorie-adequate diet, and average life expectancy at birth is only 49 years, less than two-thirds the developed world's average. Caught in the struggle to survive and trapped by ongoing cycles of hunger, malnutrition, disease, and exploitation, people have little time to worry about environmental trends. In many ways, however, they are more directly affected by environmental quality and environmental problems than those of us in the industrial world, because most of the rural poor gain their livelihood directly from the soils, waters, grasslands, forests, and wildlife, that are generating international concern. The combined effects of failed rains, population growth, and poor land management have resulted in great human tragedy.

The example we use in this chapter is drawn from recent events that occurred in African countries located on the southern fringes of the Sahara Desert. Over the past two decades, rainfall totals have been far below the amount required to sustain the existing population, and acute famine has resulted. Although the human ramifications of the drought are of great concern, we emphasize only one possible cause of the drought and provide a perspective on the ultimate importance of understanding atmospheric circulation. Far from being an item of passing academic interest, it concerns most people of the world.

> Across Africa's broad chest, from Senegal to Ethiopia, the worst drought of the century continues to cut a 4,000-mile [6,400-kilometer] swath of devastation. After six years of light rainfall, nearly one-third of the 51 million people who live in this band from the Atlantic to the Red Sea are threatened by starvation....The drought seems to be moving southward. The usually lush tropical forests of the northern Ivory Coast, Ghana, Togo, and Dahomey have received so little rain that their coffee and cocoa crops are far below normal....More than 3,000 elephants, lions, giraffes, and buffaloes have starved to death in Cameroon's Waza National Park [1].

So Time described the disaster in Africa in 1974. Descriptions of the suffering of local peoples were widely reported and graphically displayed on television. In this period, the tragic dimension of drought and famine became etched into the global conscience, and massive relief efforts were undertaken. Most recently, drought has ravaged Sudan and Ethiopia and claimed the lives of hundreds of thousands. Why did the drought occur? It is a difficult question to answer, for the explanations go beyond rainfall totals and into the realm of demography and human activity.

SOME HUMAN FACTORS

Before analyzing the climate of the drought area, we should examine some other prevailing viewpoints. According to some writers, the effects of the African drought have been worsened because of overpopulation and the resulting misuse of land. The consequences of this misuse, leading to desertification, are well known to most Americans from the Dust Bowl experience, and a brief summary of its history indicates the potential for disaster wherever people farm on marginally moist land.

The Dust Bowl, located in the southern part of the Great

Plains, has always experienced periodic droughts. Such droughts caused little lasting damage on land that was covered with natural grasses adapted to periodic dryness. However, with the settlement of the Great Plains and the plowing of the land for agriculture, the situation changed. In 1880, the first wave of settlers arrived in the area, at a time when rainfall was sufficient for raising crops. Many acres of land were plowed up. By 1890 a drought had set in, and the farmers, unable to earn a living from the land, deserted their farms and moved west. At the turn of the century, wetter conditions returned, and another wave of settlers plowed the land. These farmers also eventually failed, for in 1910 another drought occurred. Again rain returned, and in the 1920s millions of acres of the natural grass cover were plowed and planted with wheat. The inevitable drought came again in the 1930's. This time, enormous areas were bare of grass or crops. The exposed soil, dried and cracked from lack of water, broke up into small particles that were swept up by the wind. Enormous dust storms resulted as the former topsoil, with nothing growing in it to hold it in place, was blown away. In many locations, between 5-30 centimeters (2-12 inches) of the topsoil were lost. When the dust settled, it covered roads, fences, cars, and even farmhouses. Loss of the topsoil ruined the farming and led to a mass migration from the stricken area. In his novel, The Grapes of Wrath, John Steinbeck wrote eloquently about the desperate migration of thousands who sought a new life in California.

Although the droughts and the wind were immediately responsible for soil erosion and creation of the Dust Bowl, the real reason for the disaster was the failure both to preserve and protect the ground cover to prevent erosion of cultivated areas from the effects of wind. There has been similar misuse of land in the Sahel, which is an Arabic word meaning border or coast and describes the southern fringes of the Sahara Desert, extending from Senegal to Ethiopia (Figure 4-1).

Figure 4-1. The location of the Sahel in Africa.

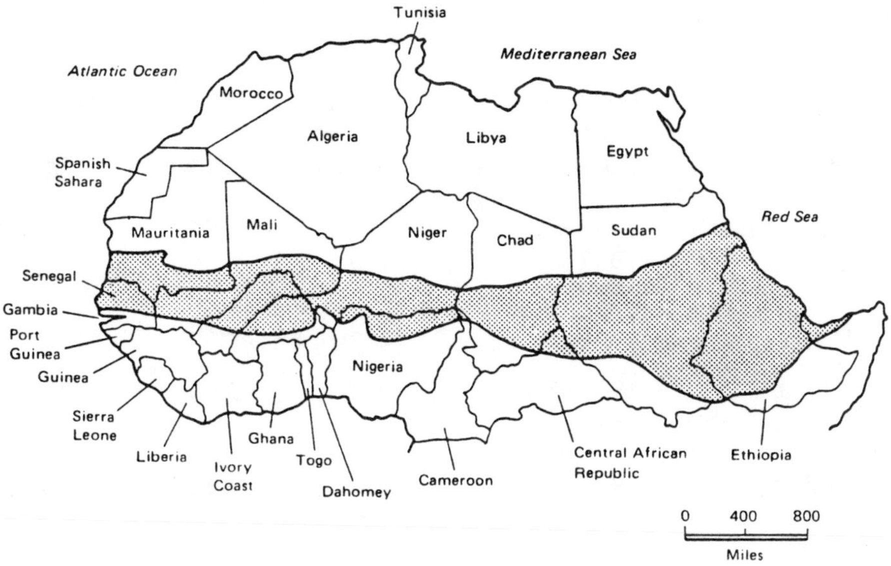

Like the Dust Bowl, this area has also experienced periodic droughts throughout history, and soil misuse has contributed to the current acute conditions. It typically receives rainfall during only four months of the year, and this rainfall is characterized by a high degree of variability, both temporal and spatial (Table 4-1). The ecosystem, however, was sufficiently strong to sustain herds of cattle, forest activity, and crops of millet and sorghum, as long as the resources were not destroyed.

The colonial quest for raw materials and the post-colonial emphasis on export crops led to an expansion of rain-fed agriculture. As the area of settled farms increased, nomadic herders were compressed into increasingly smaller areas with less reliable rainfall. Many of the land-use changes were motivated by economic and political expediency with little

Table 4-1. Precipitation Characteristics for Selected Sahel Stations, Measured in Millimeters.

Weather Station	Country	Period of Record	Average Annual Rainfall for Period	Range of Annual Totals High	Range of Annual Totals Low
Atar	Mauritania	1935-36 1941	126	259	15
Tessalit	Mali	1951	89	178	29
Agades	Niger	1930-36 1941	142	285	37
Khartoum	Sudan	1903	156	382	26

concern for understanding or anticipating the environmental or social consequences.

In the 1950's and early 1960's, greater than normal rainfall, combined with the expansion and intensification of agriculture, led to increasing human and animal pressure on the land. In 1970 the six Saharan countries of Senegal, Mauritania, Mali, Upper Volta, Niger, and Chad had some 24 million people and about the same number of animals. This was one-third more people and twice as many animals as the land was supporting under more nearly optimal conditions in 1930. When the drought did occur, crop failure meant less food for the existing population. Crops that were normally used as seed for the following year were consumed and thus the effects of the food shortage were extended. At the same time, the parched grasses were unable to support the large number of animals, and competition for the meager growth that did occur resulted in overgrazing and destruction of the land. Without a vegetation cover, the soil dried and cracked, and the process of desertification was underway (Figure 4-2).

Figure 4-2. People of all ages became refugees from the drought-stricken area. (Courtesy of United Nations).

The Sahel has an environment that can sustain only cautious use; and when the rains failed, the equilibrium was destroyed. While the intricate social and economic conditions added to the magnitude of human suffering, drought was certainly the base of the problem. Since the cause and the extent of the drought are related to the modified patterns in the prevailing weather systems, we need to understand the synoptic climatology of the Sahel.

THE CLIMATIC BACKGROUND

The Sahel is located in a zone where rainfall varies enormously from year to year (Table 4-1). A good working model explaining rainfall distribution, formulated by H.O. Walter [5], is illustrated in Figure 4-3. The annual cycle is dominated by the northward advance, and retreat of warm, humid air associated with the equatorial low pressure system variously called the intertropical front (ITF), the intertropical convergence (ITC), and the intertropical discontinuity (ITD). Here, we shall refer to it as the ITD. Figure 4-3a, shows the position of the northernmost advance of this pressure belt. In Zone A, the weather is dominated by Saharan air of low humidity. To the south of Zone A, the zones are grouped as follows (see Figure 4-4):

Zone B: There is a discontinuity here between the front and the hazy Saharan air of Zone A. Although this is the leading edge of the advancing air mass, there is little cloud development or rain.

Zone C: In this zone, there is more frequent cloud cover and showers and thunderstorms occur.

Zone D: This zone is characterized by fewer showers and more clouds of the stratus type. Rainfall is more prolonged.

Zone E: This area is largely associated with coastal regions where heavy cloud cover occurs, but where periods of rain are not frequent.

Figure 4-3. The Walker Model, showing the seasonal location of the ITD and related weather zones.

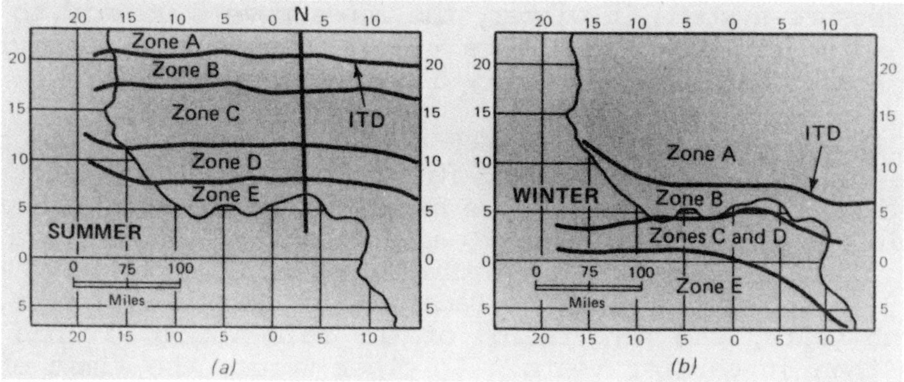

Figure 4-4. A north-south cross section through the atmosphere during the maximum summer advance of the ITD.

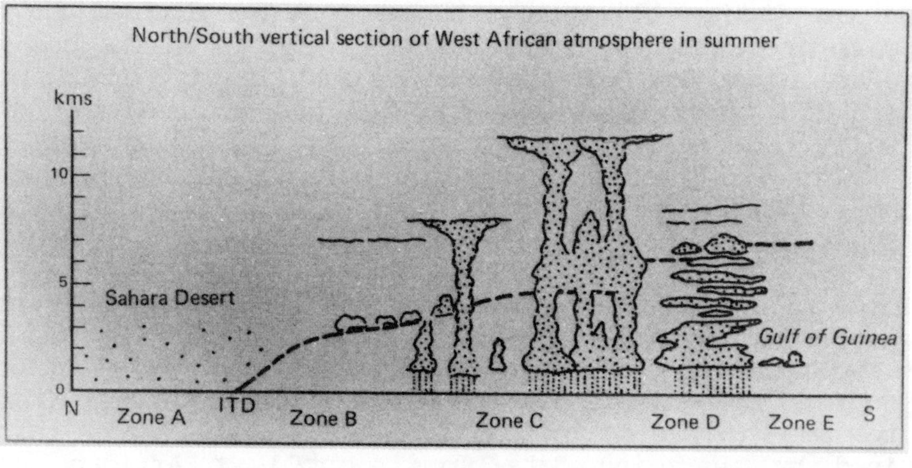

The zone locations in Figure 4-3a occur during the summer when the sun radiates the greatest amount of energy in the Northern Hemisphere, which, in turn, causes pressure systems to migrate to their most northerly location. Following the retreat of pressure systems in winter, the zones move southward to give the situation shown in Figure 4-3b. The Saharan fringes of Zone A are dominated by the dry desert air.

Although this model does not answer all questions about the West African rainfall pattern, it does provide a good picture of why the Sahelian countries have highly varied seasonal rainfall. Should the migration of the ITD advance only 1° of latitude less than normal, large areas would experience much less rain. This, in effect, is what the Sahelian drought is all about. In recent years, the penetration of the rain-making air has been less than in earlier years. In other words, the limit of the dry Sahara climate appears to be moving further south. Why the invasion of moist air has been restricted in recent years is a question that has fascinated climatologists and led to some interesting theories.

THE CAUSE OF THE DROUGHT

On the whole, climatologists seem to be divided into two camps on the cause of the Sahelian drought. On the one hand, there are those who feel that the drought is part of the normal pattern of events; it is an event that has occurred in the past and will occur again in the future. On the other hand, there are others who feel that the area is undergoing progressive drying and that the drought is part of a "desertification" of the Sahel.

Those climatologists who suggest that the drought is normal draw upon varied evidence to back their claims. By examining just two types of evidence, we can easily see the fluctuations they cite. Table 4-2 shows the varying levels of Lake Chad and the maximum-minimum deviations from the long-term mean rainfall obtained by averaging data from eight West African weather stations. Interpretation of the data suggests that since 1900 there have been at least three above-normal wet periods and three below-normal dry periods.

Table 4-2. Maximum and Minimum Rainfall Periods in West Africa (1874-1971).

(a) Average Data for 8 West African Stations[1]		(b) Comparative Water Levels in Lake Chad
Wet Periods Maximum Year	Dry Periods Minimum Year	
		High level, 1874
	1913	Low level, 1908-1914
1920		High level, 1916
1931		
	1942	Low level, 1940-1946
1957		High level, 1950-1962
	1971	

[1]Kano, Sokoto, Miduguri, Nizmey, Zinder, Segou, Maradi, and Lort Lamy.

Such climatic and hydrologic evidence indicates that periodic droughts are an integral part of the climate of the Sahel. Indirectly, the evidence also suggests that the relative severity of the current drought may well be a result of misuse of resources through overpopulation of animals and people. There is growing evidence that changes in surface albedo and reduced evapotranspiration, both associated with the destruction of protective vegetation, may be related to reduced precipitation in the region.

Other climatological analysis of the cause of the drought has less data to support it but provides a much more frightening prospect. This explanation again draws upon the seasonal advance of the ITD but suggests that the current picture is part of a worldwide change in climate. Perhaps the leading exponent of this point of view is Professor Reid Bryson of the University of Wisconsin. His argument is based on the idea that the atmosphere is a huge system and that if part of the system is modified, changes will occur elsewhere. Because the Sahel is part of a global climatic system, it must be treated as such if it is to be understood.

BRYSON'S ANALYSIS

Studies of the general circulation of the atmosphere have shown that two basic climatic regimes exist. From the equator to approximately the thirtieth parallel north and south, the Hadley Regime dominates. Essentially, this regime is a giant convective cycle where warm air rises at the equator, flows poleward aloft, and eventually subsides in subtropical areas, at a latitude of about 30°. At the equator the rising air is associated with a low pressure system; the subtropical areas of subsidence are high pressure belts. A representation of this regime, as it occurs in the Northern Hemisphere, is shown in Figure 4-5a.

Poleward of the Hadley Regime, circulation patterns are dominated by the Rossby Regime, and Figure 4-5b shows this circulation as it occurs in the Northern Hemisphere. The surface weather patterns associated with the Rossby Regime are highly changeable, with high and low pressure systems traveling across the middle latitude areas. The constantly changing conditions are related to the upper air currents shown in Figure 4-5b. Principally, these currents consist of a global, westerly flow of air in which large undulations occur. It is the size and location of these undulations, or Rossby Waves, that basically determine the type of weather experienced at the surface. As the diagram shows, the leading edge of the upper air waves is characterized by a rapidly moving stream of air, a jet stream.

Figure 4-5. a) The Hadley Regime shows air rising at the equator and subsiding in subtropical latitudes. b) The Rossby Regime comprises a westerly flow of upper air currents that flow in large meandering motions. c) A composite of the two regimes shows the subtropical high pressure cells located between the two circulation regimes. (From The Weather Machine, by Nigel Calder. Reprinted by permission of The Viking Press and the British Broadcasting Corporation.)

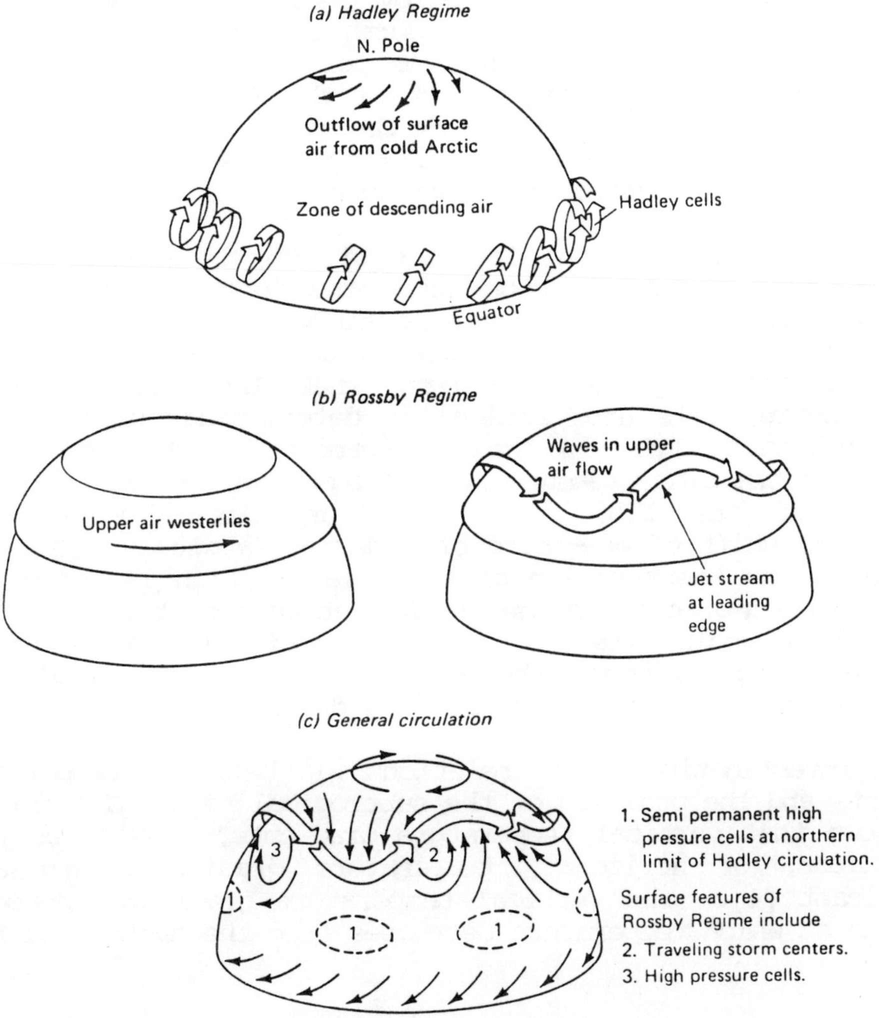

The combined regimes are illustrated in Figure 4-5c. The limit of the most southerly extension of the Rossby Regime is made up of the high pressure cells that are formed by the descending air of the Hadley Regime. The actual location of the junction between the regimes is partially determined by the difference in temperature--the temperature gradient--between the equator and the poles. In the Northern Hemisphere the greatest temperature gradient occurs in winter, when pole areas are intensely cold but tropical temperatures are only slightly lower than those of the warm season. The resulting large temperature gradient causes the Rossby Regime to dominate and extend toward the equator, displacing the Hadley Regime. In summer, the temperature gradient between the North Pole and the equator is less; so the Rossby Regime migrates poleward, and the tropical circulation occupies a larger part of the Northern Hemisphere. Since the high pressure cells mark the junction between the two regimes, these cells are seen to migrate with the seasonal variation of pole to equator temperature gradient.

The significance of the location of the two regimes to the Sahelian drought depends upon where the junction--the desert-producing high pressure cell--is located. When the cells move poleward, the belt of low pressure at the center of the Hadley Regime--the ITD--migrates with them. Thus, the positions of the high pressure cells are critical in determining the extent of the rain-bearing ITD. A small difference in the latitudinal location of the high pressure systems has a marked effect on the position of the ITD. In fact, an almost unobservable latitudinal shift of one-third of a degree (about 35 kilometers (22 miles)) in the position of the Atlantic High is associated with a whole degree of change in the latitude of the ITD. Such a difference would affect a large strip of land in the Sahel, and this strip, without the advance of the ITD, would not receive rain and would be dominated by desert conditions.

In investigating the relationship between temperature gradients and the position of the subtropical highs, Smagorinsky has shown that vertical temperature gradients, as well as pole to equator, or horizonal, temperature gradients, are also significant [4]. The vertical temperature gradient represents the rate at which temperature decreases from the earth's surface upward into the atmosphere. He shows that one-third of a degree of latitudinal change in the position of the Atlantic High, which, as noted earlier, represents a one degree change in latitude in the position of the ITD, can be caused by an increase of vertical temperature change of 0.06°C for every kilometer (about 0.17°F per mile) of ascent.

Drawing on Smagorinsky's finding, Bryson investigated the way atmospheric pollution in the middle latitudes might affect the vertical temperature gradient. Carbon dioxide, introduced into the atmosphere by the use of fossil fuels in urban, industrial activity, contributes toward an enhanced greenhouse effect, and, therefore, average temperatures in middle latitude, industrial areas should increase. But they are decreasing. Bryson postulated that carbon dioxide warms only the surface temperature and does not change temperatures at higher levels of the atmosphere. However, such a modification affects the vertical temperature gradient, and, as Smagorinsky has shown, influences the location of the subtropical highs.

Atmospheric pollution also affects the horizontal temperature gradient. Particles in the atmosphere reduce sunlight intensity and thus contribute toward temperature decline. Machta has shown that most of the increase in particles is at high latitudes [2]. This finding is in accord with the research of Reitan, who has determined that the decline in average temperatures in the Northern Hemisphere since 1945 has been in high latitudes, with scarcely any change in tropical temperature [3]. Bryson attributes much of the increase in turbidity (density of atmospheric particles) to human activities. He contends that 17 percent of the temperature variance caused by turbidity over the last century is a direct result of particles that we have added to the air. He further suggests that in recent decades, increased additions may account for 30 percent of the variance. Therefore, atmospheric pollution in the middle latitudes affects both the horizontal and vertical temperature gradients, which, in turn, affect the position of the subtropical highs and the northward penetration of the ITD.

We can see that Bryson's argument, while controversial, has a basic logic to it. By relating the advance of the ITD to circulation controls, he establishes a mechanism for producing the Sahel drought. Furthermore, he shows that the modification is related to people's misuse of the atmospheric reservoir in areas far removed from the site that has experienced the effects of the change. It is a challenging notion and one that needs to be pondered.

CONCLUSION

For the past two decades, areas on the margin of the Sahara have experienced profound human suffering. Initially, in the early 1970's the focus of concern was Genegal and the Western Sahelian countries. While the rains returned to some areas, they have continued to falter in others. Most recently the effects of drought have impacted the Sudan and Ethiopia. There is reason to believe that if human and animal pressure on the land had been less severe, the effects of the drought would have been neither as dramatic nor as swift.

The challenge, then, is to implement long-term rehabilitation strategies based on the capabilities of the region's resources, sound management of these resources, and consideration of traditional land-use practices. No one knows how much it will rain, and we are still striving to gain a more complete understanding of Sahelian climate. Some climatologists consider the drought to be merely a passing phase to be recovered from. Others are concerned that the environmental degradation and loss of vegetation and soil are encouraging the spread of aridity in a process of desertification that may be irreversible. Still others see this recent drought as part of long term climatic change. This will be explored in Chapter 7. There is no happy conclusion to this chapter; there are just too many unknowns.

Persistent drought reinforces the need for a change in the relationship of people and the land which provides their livelihood. Even from the most positive point of view it is evident that drought will occur again. But we can hope that people will use the land more carefully in the future, so that potential devastation is diminished.

SUGGESTIONS FOR FURTHER WORK

1. In recent years, drought in various parts of the world has made newspaper headlines. While you may not live within an area that has actually experienced a drought, its effects may well influence you. Explain why this is true, or if you disagree, why it is not true.
2. From recent newspapers and news magazines, collect as many items as you can about drought. Such items frequently provide "human interest" accounts of the response of individuals to drought. Do there appear to be any common themes in people's reactions, or are there clear "national" responses?

SELECTED SOURCES FOR FURTHER READING

Gritzner, Jeffrey Allman. The West African Sahel: Human Agency and Environmental Change. Chicago: University of Chicago Geography Research Paper, No. 226, 1988.

National Research Council, Advisory Committee on the Sahel. Environmental Change in the West African Sahel. Washington D.C.: National Academy Press, 1983.

Oliver, J.E. Climate and Man's Environment. New York: Wiley, 1973.

Scott, Earl. Life Before the Drought. Boston: Allen & Unwin, 1984.

Somerville, Carolyn M. Drought and Aid in the Sahel: A Decade of Development Cooperation. London: Westview Press, 1986.

Winstanley, Derek. "Africa in Drought: A Change of Climate?" Weatherwise. Vol. 38, (April, 1985), pp. 15-81.

REFERENCES

1. "A Feast for Vultures," Time, April 8, 1974.
2. Machta, Lester. "Mauna Loa and Global Trends in Air Quality," Bulletin American Meteorological Society. 5 (1972), pp. 402-21.
3. Reitain, C.H. "An Assessment of the Role of Volcanic Dust in Determining Modern Changes in the Temperature of the Northern Hemisphere." Ph.D. thesis. Madison, University of Wisconsin, 1971.
4. Smagorinsky, J. "General Circulation Experiments with the Primitive Equations, I: The Basic Experiment," Monthly Weather Review. 91 (1963), pp. 99-164.
5. Trewartha, G.T. The Earth's Problem Climates. Madison: University of Wisconsin Press, 1961.

Chapter 5

Severe Weather

The Context

Atmoshperic disturbances vary in origin, intensity, location, and rate of movement. Some disturbances are relatively common and have long been known and studied. Others are relatively rare, and their origins and characteristics are not well understood. Some, such as the middle-latitude cyclone that plays such an importnat role in causing the erratic, day-to-day changes in our weather, can be observed in isobar and wind patterns on the daily surface weather maps often included in newspapers and on television.

Others, such as the tornado, are much too small to be seen in the features of these maps. Because of their great catastrophic potential, severe storms create a fascination and interest that ordinary weather events cannot provide. Death, destruction, and dislocation often accompany violent storms.

On a sultry summer afternoon in 1928, Will Keller was tending his fields when he saw the twisting tentacle of three tornadoes approaching rapidly. Calling to his family, Will gathered the others and ran to the cyclone cellar. Just as he was about to close the door, he paused to take a quick look and saw the end of the funnel begin to rise. In comparative safety he watched the funnel, and this is what he described.

> Two of the tornadoes were some distance away and looked like great ropes dangling from the parent cloud, but the one nearest was shaped more like a funnel, with ragged clouds surrounding it. It appeared larger than the others and occupied the central position, with great cumulus clouds over it.
>
> Steadily the cloud came on, the end gradually rising above the ground. I probably stood there only a few seconds, but was so impressed with the sight it seemed like a long time. At last the great shaggy end of the funnel hung directly overhead. Everything was still as death. There was a strong, gassy odor, and it seemed as though I could not breathe. There was a screaming, hissing sound coming directly from the end of the funnel. I looked up, and to my astonishment I saw right into the heart of the tornado. There was a circular opening in the center of the funnel, about fifty to one hundred feet [15-30 meters] in diameter and extending straight upward for a distance of at least half a mile [.8 kilometer], as best I could judge under the circumstances. The walls of this opening were rotating clouds and the whole was brilliantly lighted with constant flashes of lightning which zigzagged from side to side....
>
> Around the rim of the great vortex, small tornadoes were

constantly forming and breaking away. These looked like tails as they writhed their way around the funnel. It was these that made the hissing sound. I noticed the rotation of the great whirl was counterclockwise, but some of the small twisters rotated clockwise....The tornado was not traveling at a great speed. I had plenty of time to get a good view of the whole thing, inside and out [3:205-206].

This account represents a typical description of a tornado, except that funnels are usually single rather than multiple in form. Keller was one of the few people to have looked directly into the eye of a tornado who lived to tell about it.

Others were not so fortunate. On March 18, 1925, beginning in the afternoon at 12:55 p.m., the infamous tri-state tornado ravaged the countryside from Reynolds County in southeastern Missouri across southern Illinois to Pike County in southwestern Indiana, where it finally dissipated nearly four hours later at 4:30 p.m. The tornado killed 695 people, injured 2,027 and caused 16.5 million dollars worth of property loss, in 1925 dollars. Seven other tornadoes in the region that day added significantly to the property damage and brought the total life loss to 740 [1:91].

On April 3 and 4, 1974, the largest known tornado outbreak in the nation's history occurred. This super outbreak generated 148 tornadoes and lasted just over 16 hours. In that period 315 individuals were killed, nearly 5,500 were injured, and property damage amounted to over 600 million dollars. In recent years, property damage averages between 200 and 300 hundred million dollars annually [4].

Although small in size and localized in effect, no other atmospheric disturbance is as dramatic or as destructive as the tornado. Thus, these storms have generated considerable scientific and public attention.

Although different theories on tornado genesis have been proposed, we, as yet, are not in a position to give definitive answers. In recent years, however, new tools such as Doppler radar and satellite data, have given us a much clearer picture of the severe convective events that are most commonly associated with tornadoes.

THUNDERSTORMS

Before discussing tornadoes, we must first understand the life cycle of the thunderstorm which spawns the more destructive tornado and other severe weather such as hail. Thunderstorms are most common where the ingredients of high humidity, high temperatures, and high instability are present. It is not surprising that equatorial regions have the greatest number of thunderstorms and that poleward of 60° latitude the storms are virtually unknown. In the United States, thunderstorms are most common in spring and summer and have their highest frequency along the Gulf Coast and Florida (Figure 5-1). The presence of stable anticyclonic air along the Pacific coast accounts for the low frequency in that area of the country.

Throughout the years, 45,000 thunderstorms buffet the earth each day; and while the vast majority are not severe, a few are capable of generating violent activity. Thunderstorms are generated by temperature imbalances in the atmosphere and represent the well-developed processes of convection. Convection is a process in which heat is transferred through the vertical rise of parcels of low-density air. In the convective process, warmer surface air is exchanged with the cooler air above it. Since the atmosphere does not conduct heat well and is transparent to incoming solar radiation (insolation), heat tends to build up at lower levels as the earth's surface is warmed and as cooling of the air above takes place.

This process -- the initial stage of the development of a thunderstorm -- places warmer, lighter air below more dense, colder air and creates a condition of instability. As a result, the air layers turn over, with the warm air rising rapidly and the cooler, denser air sinking. These bubbles of air move through the atmosphere much like water boils on a stove. Although less obvious, cool air flowing into upper reaches of the atmosphere can produce a similar situation.

Figure 5-1. Average Number of Thunderstorms Each Year in the United States.

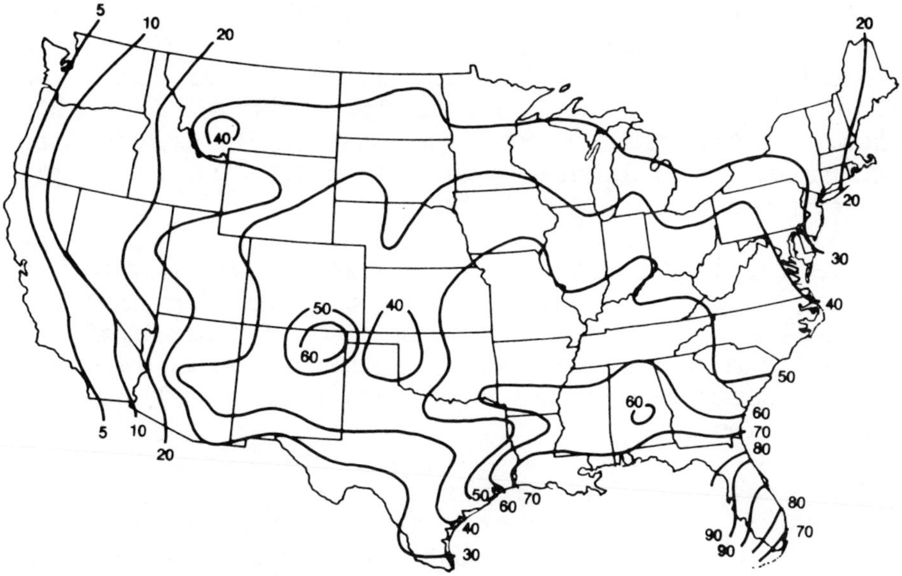

Mechanical conditions may also initiate convective activity. Air may be forced upward by the invasion of a cold air mass or along mountain slopes, which create unstable buoyancy.

In the initial stage of development, vertical motion carries warm air to a level in the atmosphere where saturation occurs and visible clouds begin to form. Continued motion produces the towers or domes, known as cumulus clouds. Frequently, strong wind flow above the clouds enhances this convective process.

Eventually, the updraft reaches a level where water vapor changes in form to liquid or ice spicules, and the release of energy in this process becomes the major source of energy to keep the cloud growing. The cloud particles grow by coalescence and collision. Ultimately, the weight of these moisture particles becomes too great for the updraft to sustain, and drops of water or pellets of ice begin to fall. Their drag creates a downdraft, which is often manifested by precipitation and gusts of wind near the earth.

In this stage, the downdraft is further aided by the influx of cool, dry air surrounding the cloud in a process called entrainment. This marks the beginning of the storm's mature stage. During this stage, downdrafts exist side by side with updrafts. As the cloud reaches the top of the unstable region, the updrafts spread laterally and, coupled with the high level winds, form the familiar anvil head.

Eventually, downdrafts dominate the cloud and initiate the dissipating stage. The cooling effect of falling precipitation along with the downdraft and influx of colder air aloft, signal the end of the storm's activity. Precipitation decreases and downdraft ceases. Without a supply of moisture, the storm cell spreads and dies.

We have described the updraft-downdraft that constitutes a single cumulonimbus cell. The life span of the cell is short, generally less than an hour, but most storms are composed of several cells. Because new cells form, it is possible for storms to last several hours.

Generally, thunderstorms progress through the three stages described. As noted, the initial development is the cumulus stage in which updrafts dominate cloud growth. the mature second stage, in which both updrafts and downdrafts are present, is often characterized by severe weather, lightning, heavy rains, and gusty winds. In the final stage of growth, a mature storm may be several miles across the base and tower to elevations of 12,000 meters (40,000 feet) or more. Finally, in the dissipating stage, downdrafts dominate and entrainment causes evaporation which leads to the ultimate death of the storm.

Isolated thunderstorms are locally impressive but not particularly dangerous, and unless someone is struck by lightning, these storms do a good deal more good than harm. The cool air that precedes the storm provides relief from the hot temperatures on a summer day, and the rain is of vital importance to farmers during the growing season.

SUPERCELL THUNDERSTORMS

Some storms, because of their large size and long duration, are known as supercells. They often occur as part of a squall line, hundreds of miles long. Squall lines most often result from the large-scale collision between cold air and warm air. As the dense cold front underrides the warm air, convection is initiated. In some cases, convection may occur in prefrontal waves, well in advance of the front. The squall line as seen from the ground is often preceded by dark cloud rolls that have downward pouches called mammitus sky. A sharp shift in wind direction, cooler temperatures, and gusty wind conditions capable of inflicting heavy damage are also common to squall disturbances. Although the squall line may continue for several hours, most of these storms are of short duration; however, they may contain the rare supercell capable of generating extremely violent weather. Like other thunderstorms, the supercell draws in warm humid air from the adjacent countryside and forces it aloft. In ordinary storms the condensing moisture produces a downdraft and ultimately the storm collapses. In a supercell the updrafts and downdrafts are separate and reinforce one another, giving the cell a long lifetime. Often strong wind shear is present, and this causes the storm to rotate, producing a mesocyclone. The cyclonic rotation may simulate a tube extending from the lower portion of the storm up to the vicinity of the jet stream region. Perhaps this cyclone action may be part of the mechanism in the formation of tornadoes, if this circulation extends groundward.

Supercells account for only a small portion of all severe weather-producing events, but the extreme nature of the severe weather associated with these storms causes a much greater percentage of injuries and damage than their number might indicate. Virtually all of the most violent tornadoes are spawned by the supercell. The remaining portion of this chapter will be devoted to the offspring of these storms. The destructive progeny of these storms are lightning, flash floods, hail, downbursts, and tornadoes.

LIGHTNING

In most years, lightning kills more people than any other circumstance of weather. In the United States each year, lightning kills approximately 100 people, injures nearly 250 people, and causes property damage estimated in the hundreds of millions of dollars. Because it usually causes an individual, rather than a group disaster, lightning tends to receive much less publicity than other severe weather events. Most fatalities occur in the open, and dangerous locations to be avoided are the high ground in a particular area, tall isolated trees, wire fences, and tractors with metal implements in the ground. Indoors, unsafe locations are near metal appliances, metal water pipes, or telephones. Two out of three people struck by lightning are not killed, and quick cardiopulmonary resuscitation can save most victims.

Lightning is produced by thunderstorms. As positive charges are carried to the top of the storm, negative charges accumulate at the base. This process produces an intense electrical field within the cloud. The growing negative charge in the lower part of the clouds attracts a growing positive charge on the earth's surface immediately below. Because air is a poor conductor of electricity, it insulates the ground charge preventing a flow of current until huge electrical differentials are built up. The electrical potential in these cases can be as high as 100 million volts. The lightning stroke occurs when the electrical potential -- the difference between positive and negative charges -- becomes large enough to overcome the resistance of the insulating air and forces the conductive path of the current to flow between the two charges. Lightning is exchanged within the same cloud; between adjacent clouds; from cloud to ground; or in some cases, where high structures are involved, from ground to cloud.

The lightning stroke produces instantaneous expansion of the air and creates a shock wave that we hear as thunder. If lightning is near-by the thunder sounds like a sharp crack; however, strokes at a greater distance produce a rumbling and growling noise. Lightning is seen at essentially the same instant it occurs because it travels at the speed of light. Thunder travels at the slower speed of sound, so we see the lightning bolt before the sound reaches us. Thus, we estimate the distance of a lightning stroke by timing the interval between sight and sound. An interval of five seconds indicates that the lightning flash is about a mile away. When a thunderstorm threatens, it is wise to be prudent because lightning poses a serious hazard.

FLOODS

Another product of thunderstorms is convective precipitation. Convective precipitation falls in the form of brief, torrential, local showers. Because convective rainfall comes in heavy showers, only a small percentage soaks into the soil. Most of the moisture disappears as surface run-off, capable of causing serious gullying and washing on plowed fields, as well as intense local flooding. Flash-floods, produced by individual thunderstorms, are localized and therefore more difficult to predict than other types of floods. In 1976, a flash-flood from an individual thunderstorm was responsible for the deaths of about 200 people, and caused more than 50 million dollars worth of property damage in the Big Thompson Canyon of Colorado. Annually, flash-floods have taken an average of more than 150 lives per year in the United States.

HAIL

Another child of the thunderstorm is hail. Hail storms are responsible for considerable damage to agricultural crops, but can also cause heavy damage to automobiles, aircraft, and rooms and windows in buildings. Relatively few people have been killed by hail in the United States; however, in April 1888, near Delhi, India, 246 people were reportedly killed during a severe hail storm. Annually, the damage to crops in the United States amounts to a total of nearly one billion dollars.

Hail, produced in central updrafts of the thunderstorm, is the most spectacular of the major precipitation forms. Hail consists of relatively large, rounded pieces of ice, which typically have a layered structure. Individual hail stones range from pea size to the size of a grapefruit. Hail stones form in a storm as the strong updrafts and downdrafts carry ice pellets alternately above and below the freezing level in the clouds. As the hail stone is buffeted about, it gathers layer upon layer of ice from the super-cooled water droplets that freeze upon impact. Finally, when it becomes so heavy that even the strongest updraft cannot sustain it, the mature hail stone plummets to earth.

Surprisingly, hail frequency and thunderstorm frequency do not show the same distribution. In the tropics, with the highest frequency of thunderstorms, hail is practically unknown. In the United States, hail rarely occurs in the warmer subtropical parts of the country such as Florida. Hail storms, strikingly local in character, are most damaging in parts of the Rocky Mountains and the Great Plains. In the High Plains area to the east of Denver, there are between 15 and 20 hail storms during May, June, and July each year (Figure 5-2). Crop losses are so common that many farmers take out insurance against hail damage.

Because of the considerable damage to crops and personal property, there is interest in hail suppression. A number of programs addressing this problem have been implemented in the United States. Most of these programs are based on injecting silver iodide into a cloud in order to create so many nuclei that competition for the liquid water supply available within a thunderstorm is created. The basic idea is that this method would result in a large number of small hail stones that either would melt during descent or would arrive as harmless tiny pellets. To date, evaluation of hail suppression experiments

Figure 5-2. Average Number of Days with Hail in the United States.

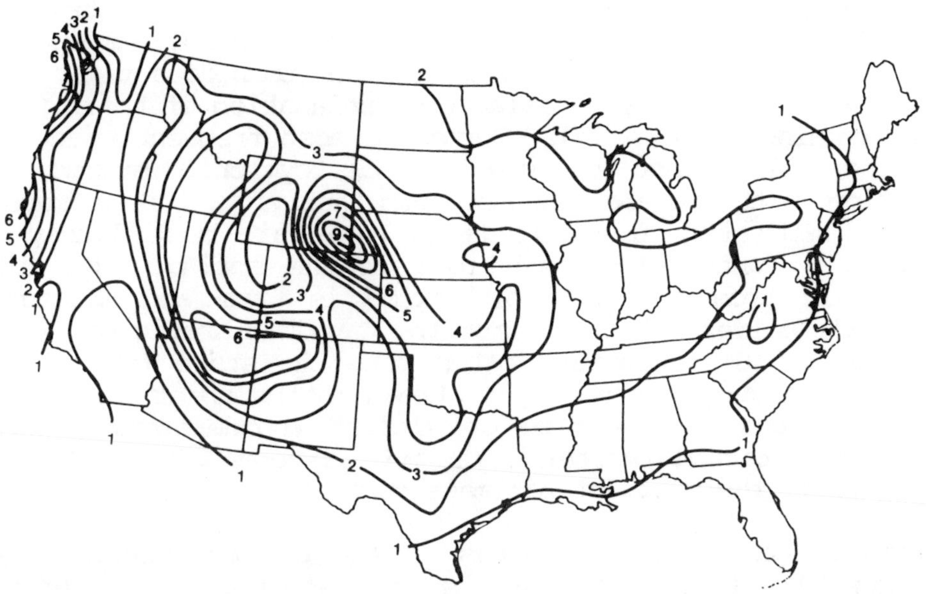

has produced mixed results. Because of the large potential benefit from such programs, it is likely that experimentation will continue.

Thunderstorms may also produce downbursts. In recent years, Fujita has called attention to downbursts that are associated with some severe storms [2:119]. These strong downdrafts are concentrations of sinking air, which fan out on striking the surface of the earth and produce damaging winds. At least some fraction of storm events, reported as tornadoes, are in-fact downbursts. Because downburst winds can be as great as those of tornadoes, it is imperative that the threat of severe thunderstorms be taken as seriously as that of tornadoes.

TORNADOES

Tornadoes, sometimes called twisters or cyclones, are Nature's ultimate storm. These local storms are of short duration but may be ranked high among the most destructive natural forces on the earth's surface. The general atmospheric conditions that are most likely to lead to tornado activity are known; however, a really satisfactory explanation of their final genesis remains to be found. Tornadoes most frequently develop along the cold front of mid-latitude cyclonic storms, in conjunction with severe thunderstorm activity. Air masses associated with the mid-latitude cyclones are most likely to have greatly contrasting conditions in the spring. Maritime tropical air from the Gulf of Mexico is warm and moisture-laden, while colder polar air from the Canadian Arctic may still be very dry. This contrast often sets the stage for intense storms. Sometimes, a strong jet stream in the upper atmosphere also contributes to the rise of violent thunderstorms, which may develop a tornado vortex. The most intense storms develop where there is the greatest air mass contrast.

Tornadoes occur in many areas of the earth's surface, but more than 90% of all such storms are reported in the United States. Because the unique physical geography of the mid-western United States provides an environment of unhindered air mass interaction, this region generates more tornadoes than any other part of the country (Figure 5-3).

Although the quixotic nature of tornadoes has led some to argue that they are as unique as fingerprints, we can comment on the general characteristics common to most tornadoes. The tornado, by definition, is a violently rotating column of air in contact with the ground. The well-known funnel shape is visible when condensation occurs or when debris is picked from the ground surface in the spiralling air. During this life-stage, changes in form occur; and ultimately, as the tornado enters its decaying stage, it may assume the shape of a rope-like tube. In some large tornadoes, the funnel shape may be absent. Sometimes, two or more tornadoes can occur within the same thunderstorm. Occasionally, in some large thunderstorms

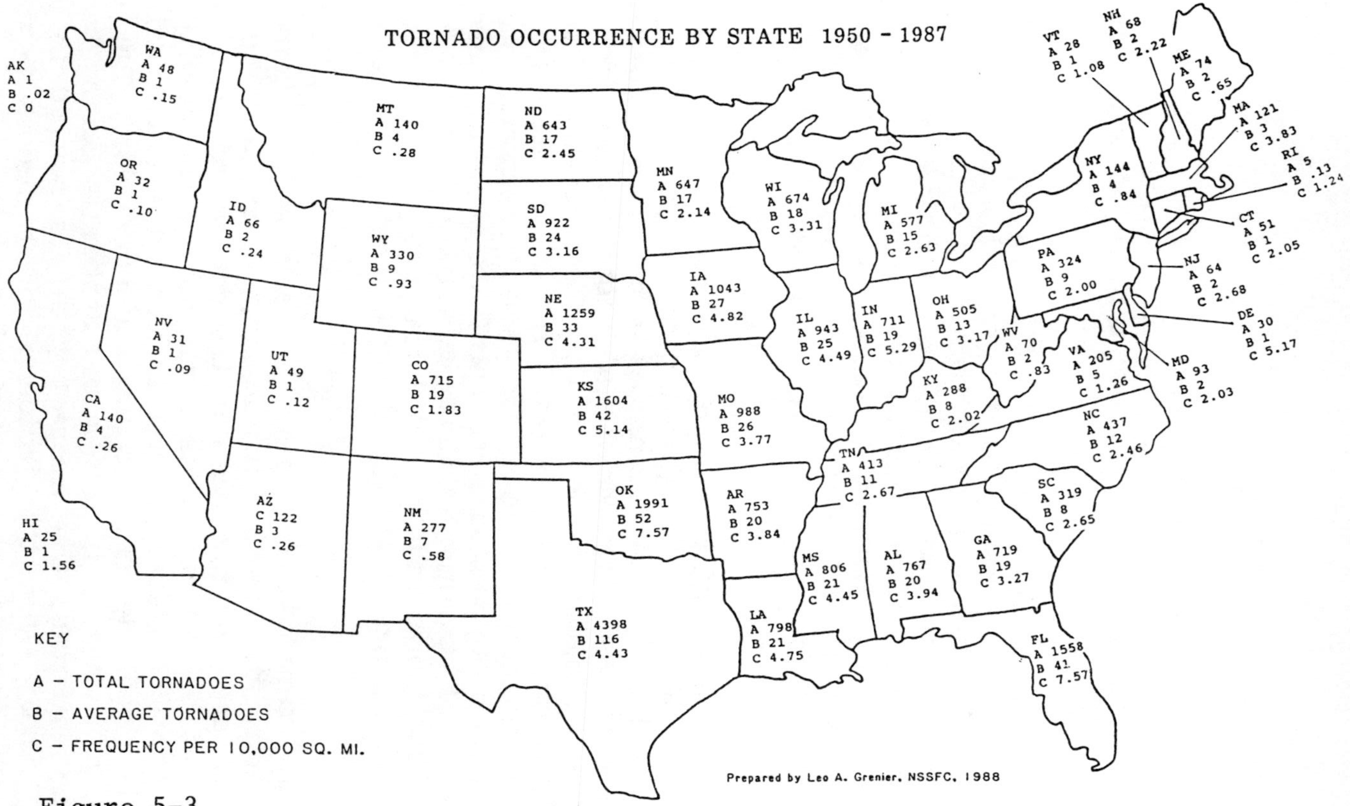
TORNADO OCCURRENCE BY STATE 1950 - 1987
AK A 1 B .02 C 0
WA A 48 B 1 C .15
OR A 32 B 1 C .10
CA A 140 B 4 C .26
HI A 25 B 1 C 1.56
ID A 66 B 2 C .24
NV A 31 B 1 C .09
MT A 140 B 4 C .28
WY A 330 B 9 C .93
UT A 49 B 1 C .12
AZ C 122 B 3 C .26
CO A 715 B 19 C 1.83
NM A 277 B 7 C .58
ND A 643 B 17 C 2.45
SD A 922 B 24 C 3.16
NE A 1259 B 33 C 4.31
KS A 1604 B 42 C 5.14
OK A 1991 B 52 C 7.57
TX A 4398 B 116 C 4.43
MN A 647 B 17 C 2.14
IA A 1043 B 27 C 4.82
MO A 988 B 26 C 3.77
AR A 753 B 20 C 3.84
LA A 798 B 21 C 4.75
WI A 674 B 18 C 3.31
IL A 943 B 25 C 4.49
MI A 577 B 15 C 2.63
IN A 711 B 19 C 5.29
OH A 505 B 13 C 3.17
KY A 288 B 8 C 2.02
TN A 413 B 11 C 2.67
MS A 806 B 21 C 4.45
AL A 767 B 20 C 3.94
GA A 719 B 19 C 3.27
FL A 1558 B 41 C 7.57
SC A 319 B 8 C 2.65
NC A 437 B 12 C 2.46
VA A 205 B 5 C 1.26
WV A 70 B 2 C .83
PA A 324 B 9 C 2.00
NY A 144 B 4 C .84
VT A 28 B 1 C 1.08
NH A 68 B 2 C 2.22
ME A 74 B 2 C .65
MA A 121 B 3 C 3.83
RI A 5 B .13 C 1.24
CT A 51 B 1 C 2.05
NJ A 64 B 2 C 2.68
DE A 30 B 1 C 5.17
MD A 93 B 2 C 2.03
KEY
A – TOTAL TORNADOES
B – AVERAGE TORNADOES
C – FREQUENCY PER 10,000 SQ. MI.
Prepared by Leo A. Grenier, NSSFC, 1988

Figure 5-3.

tornadoes may be made up of a number of smaller, intense vortices that rotate about a common center. These small vortices, called suction spots, are capable of producing great damage.

Tornadoes vary greatly in size and intensity. An estimate of intensity is made by assessing damages and wind speeds on the Fujita F-scale (Table 5-1). Weak tornadoes, with wind speeds of under 180 kilometers (about 112 miles) per hour, account for 63% of all tornadoes, but they claim less than 3% of all tornado deaths. Conversely, violent tornadoes, with wind speeds from 333 to over 483 kilometers (about 207 to more than 300 miles) per hour account for only 2% of all tornadoes, but nearly 70% of tornado fatalities.

The average width of these violent storms is over 366 meters (about 400 yards), and the typical path length is 42 kilometers (about 26 miles). Some tornadoes exceed 1.5 kilometers (a mile) in width and have travelled distances as far as 483 kilometers (300 miles) on the ground. The direction of movement is usually southwest to northeast. For the nation as a whole, 87% of all tornadoes move into the northeast quadrant. Tornadoes having any westward component are extremely rare. The average speed of movement is about 48 kilometers (30 miles) per hour, but some storms have remained stationary while others have been recorded moving at speeds of 113 kilometers (70 miles) per hour.

Although tornadoes can form in any month, temperature contrast and atmospheric instability, important to their development, are most favorable in the spring and early summer. Annually, 54% of all tornadoes occur during the spring. The number of tornadoes decreases rapidly after the summer solstice, and the two cool seasons account for only 19%. The lowest frequency of tornadoes in the United States is during December and January; whereas, the greatest total is reached in April, May, and June. When the incidence of tornadoes begins to increase in February, the center of activity lies over the central gulf states. As the season progresses during May and June, the center of maximum frequency moves through the

Table 5-1. Approximate Speed Corresponding to the Fujita Intensity Scale. (Source: Weatherwise, April 1980, p. 56.)

F-Scale	Category	Approximate Speed km/hour	Approximate Speed miles/hour
0	Weak	<116	<72
1		117-180	73-112
2	Strong	181-253	113-157
3		254-332	158-206
4	Violent	333-418	207-260
5		419-512	261-318

southern plains, then to the northern plains and Great Lakes area. This pattern is related to the increasing penetration of warm, moist air from the Gulf while contrasting cool, dry air still surges from the north and northwest. The time of day when most tornadoes develop is between noon and sunset, with 60% developing during the hottest part of the day. They are least likely during the early morning hours before sunrise.

During their brief time on earth, tornadoes are capable of performing bizarre feats; and every tornado is a potential killer, capable of great destruction. They are capable of performing freakish tricks, such as leaving chickens featherless and driving fragile grain straws into telephone poles. One storm transported two people 30 meters (100 feet) aloft and then deposited them right side up in a car, with no physical damage to vehicle or humans.

The enormous destructive power of a tornado is frightening. It can destroy an entire neighborhood in minutes, leaving total destruction in its wake. At one time it was believed that the extreme low pressure region in the core of the storm caused structures to explode outward. However, recent studies indicate that the damage to buildings is caused by quite different processes. Most work of the tornado is produced by strong rotary winds, as well as by the impact of the wind-born debris picked up by the storm.

Death and injury are not correlated with the greatest number of tornado occurrences, but result from a combination of high tornado incidence and dense concentrations of people. A majority of tornadoes are weak, and it is important to keep in mind that only 2% of tornadoes are violent, but they account for 68% of all deaths (Figure 5-4).

Because tornadoes cannot be prevented, the best protection comes from being forewarned. The National Severe Storms Forecast Center, in Kansas City, monitors atmospheric conditions, gathering information from satellites, balloons, radar, and hundreds of weather stations. When atmospheric conditions are appropriate for the formation of severe storms, a WATCH is issued. A watch indicates that severe thunderstorms and possible tornadoes are likely to occur. Since all severe thunderstorms can potentially produce tornadoes, a severe thunderstorm watch does not preclude the presence of tornadoes. When a TORNADO WATCH is issued, the center indicates that conditions are favorable for both severe thunderstorms and tornadoes. Individuals living in or near the watch area should be aware of the signs of threatening weather. A TORNADO WARNING is issued when a tornado has actually been sighted in an area or when it is indicated by radar. The warnings have substantially reduced the number of deaths and injuries. Although the mathematical probability of a given location being struck by a tornado is low, tornado watches and warnings should never be taken lightly.

Some areas, however, have seemingly defied the odds. For example, Oklahoma City has been struck twenty-six times since 1892; but perhaps the most unusual record is from the small town of Codel, Kansas. This community was hit in three consecutive years--1916, 1917, and 1918--and, incredibly, each time the day was May 20th.

TORNADO DEATHS BY STATE 1950 - 1987

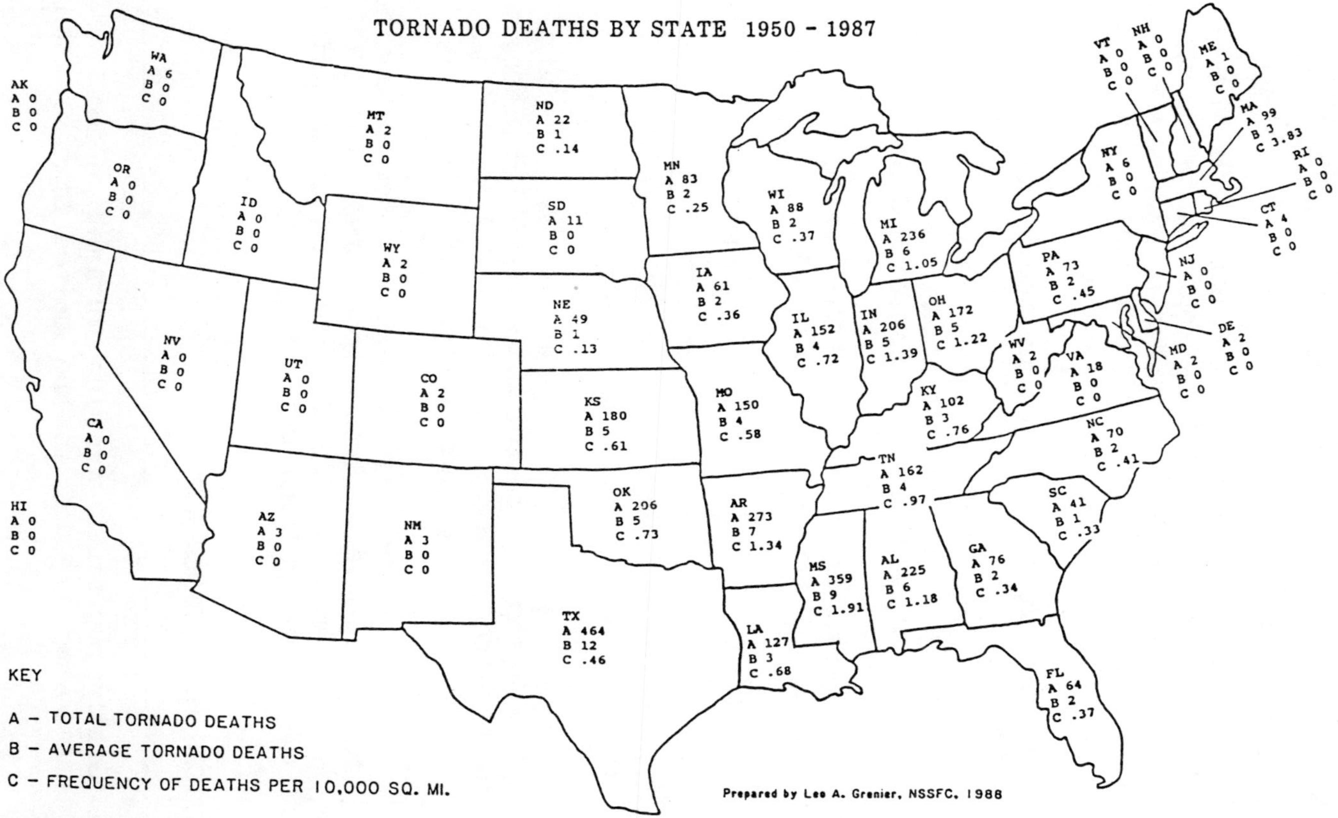

KEY

A – TOTAL TORNADO DEATHS

B – AVERAGE TORNADO DEATHS

C – FREQUENCY OF DEATHS PER 10,000 SQ. MI.

Figure 5-4.

CONCLUSION

Because of the fierce power concentrated in the narrow path of the tornado, there is little that can be done to protect property; however, the National Weather Service has issued a set of rules which, if followed, can reduce the risk of injury and death. Some of these important rules are:

1. If outdoors, lie flat in the nearest depression or ditch.

2. If in a vehicle, do not attempt to drive away from the storm, but leave the vehicle for a ditch.

3. If living in a mobile home, seek more substantial shelter, as mobile homes are particularly vulnerable.

4. If living in a home with a basement, seek refuge near the basement wall.

5. If living in a home without a basement, take cover in the smallest room with stout walls, or under heavy furniture.

6. If in a shopping center or other public place, move to a designated shelter; normally, interior hallways on the lowest floor are best [4:15].

The tornado is Nature's ultimate storm. A tornado's life is usually short; however, when a populated area is hit, the destructive path must be measured not just in dollars but also in terms of terror and human suffering.

SUGGESTIONS FOR FURTHER WORK

1. Look at the map which contains information about the number of tornado occurrences in the United States. Which five

states have had the most tornadoes? Which of the five states has had the least number of tornadoes? How many tornadoes has your home state had within the 37 year period?

2. Determine what location in your home to seek cover from a tornado. Explain your choice.
3. Explain the difference between a tornado watch and a tornado warning. What conditions must be present for each to be issued?

SELECTED SOURCES FOR FURTHER READING

Frazier, Kendrick. The Violent Face of Nature: Severe Phenomena and Natural Disasters. New York: Morrow, 1979.

Simpson, Robert H. and Herbert Riehl. The Hurricane and Its Impact. Baton Rouge: Louisiana State University Press, 1981.

Whipple, A.B.C. Storm. Alexandria, Virginia: Time-Life Books, 1982.

REFERENCES

1. Brudt, Jan. "The Tri-State Tornado". Weatherwise, Vol. 39, No. 2, April, 1986.
2. Doswell, Charles A. The Operational Meteorology of Convective Weather, Volume 2: Storm Scale Analysis. National Oceanic and Atmospheric Administration Technical Memorandum, ERL ESG-15. Boulder, Colorado: April, 1985.
3. Justice, Alonzo A. "Seeking the Inside of a Tornado". Monthly Weather Review, 5 (1939), pp. 205-206.
4. U.S. Department of Commerce, National Oceanic and Atmospheric Administration, National Weather Service. Tornado Safety. Washington: January, 1982.

Chapter 6

A Trans-Boundry Pollution Problem: Acid Rain

The Context

One out-growth of the industrial revolution has been the significant increase in acid rain. Acid rain usually originates as by-products from the consumption of fossil fuels. When fossil fuels are burned, some sulfer, nitrogen, and carbon are oxidized and released in aerosol form. These compounds are carried upward by the convection associated with heating of air by the fossil fuel combustion. The resulting particles are incorporated in water droplets during the coalescence process and return to earth as precipitation. When this occurs, the oxides mix chemically with atmospheric moisture and form weak sulfuric, nitric, and carbonic acids which are then an integral part of precipitation. The term most often applied to this acidified precipitation is "acid rain" [7].

As recently as 25 years ago, air pollution was largely considered a local problem that could be solved by building tall stacks in order to spread industrial emissions and reduce ground level concentrations. Toward the end of the 1960's, however, serious acidification of precipitation was observed over large areas of Europe, including areas far away from major industrial centers. Since then, long-range transport of air pollutants and the ecological effects of acid rain have been studied in a number of national and international research programs [2].

MEASURING ACID RAIN

Acidity of a substance is associated with the relative abundance of free hydrogen ions (H+) when the substance is in a water solution. It is measured on a logarithmic pH scale, where a value of 7 indicates neutrality: decreasing values on the scale indicate increasing acidity and increasing values represent alkalinity.

Absolutely pure water has a pH of 7, but, if left standing in clean air, its pH will decrease to near 5.6, a result of the absorption of atmospheric carbon dioxide to form weak carbonic acid. Most rain water will have this pH value; acid rain is thus considered to exist when the pH is less than 5.6 For this increased acidity to occur requires that the atmosphere contain chemicals, to provide an acid source, and that the atmosphere have the ability to deposit the material at the surface. This latter property is completed through the process of atmospheric cleansing.

The main process that results in acid deposition is the removal of chemicals through precipitation, the "wet deposition" process, starting with cloud droplet formation, and deposited as part of the eventual precipitation. Cleansing can also occur through "dry deposition". This implies that substances in the atmosphere are deposited through gravitational settling. Large particles settle under their own weight; very small particles may be carried far from their source, to eventually coagulate or polymerise such that they are large enough to fall under their own weight. Such particles are also often returned to earth through the wet depositional process.

The wet and dry depositional processes do not cover all aspects of deposition. Other forms include fog droplet interception (especially marked in high altitude forests), dew, frost and rime icing.

The current distribution of wet deposition acid rainfall in North America is shown in Figure 6-1. The eastern part of the continent, centered upon the northeastern states and southeast Canada, clearly has a higher acidity than the remainder of the area. In part, this distribution reflects the conditions in the drier west, where contributions of alkaline substances to the atmosphere alter the chemistry of precipitation.

It need be noted that the distribution shown in Figure 6-1 is based upon data derived from a variety of sources. The collection of data concerning acid precipitation is now only partially standardized, and care need be used in data interpretation.

THE NORTH AMERICAN SITUATION

In the eastern United States the major component of acid rain is sulfuric acid, comprising as much as 65 to 70 percent of the rain's acidity (Fig. 6-2). At least 1,000 acid-sensitive lakes in the eastern United States, especially in New England, are highly acidic (pH below 5.0) and have suffered sharp declines

Figure 6-1. Distribution of wet-deposition acid rainfall in North America.

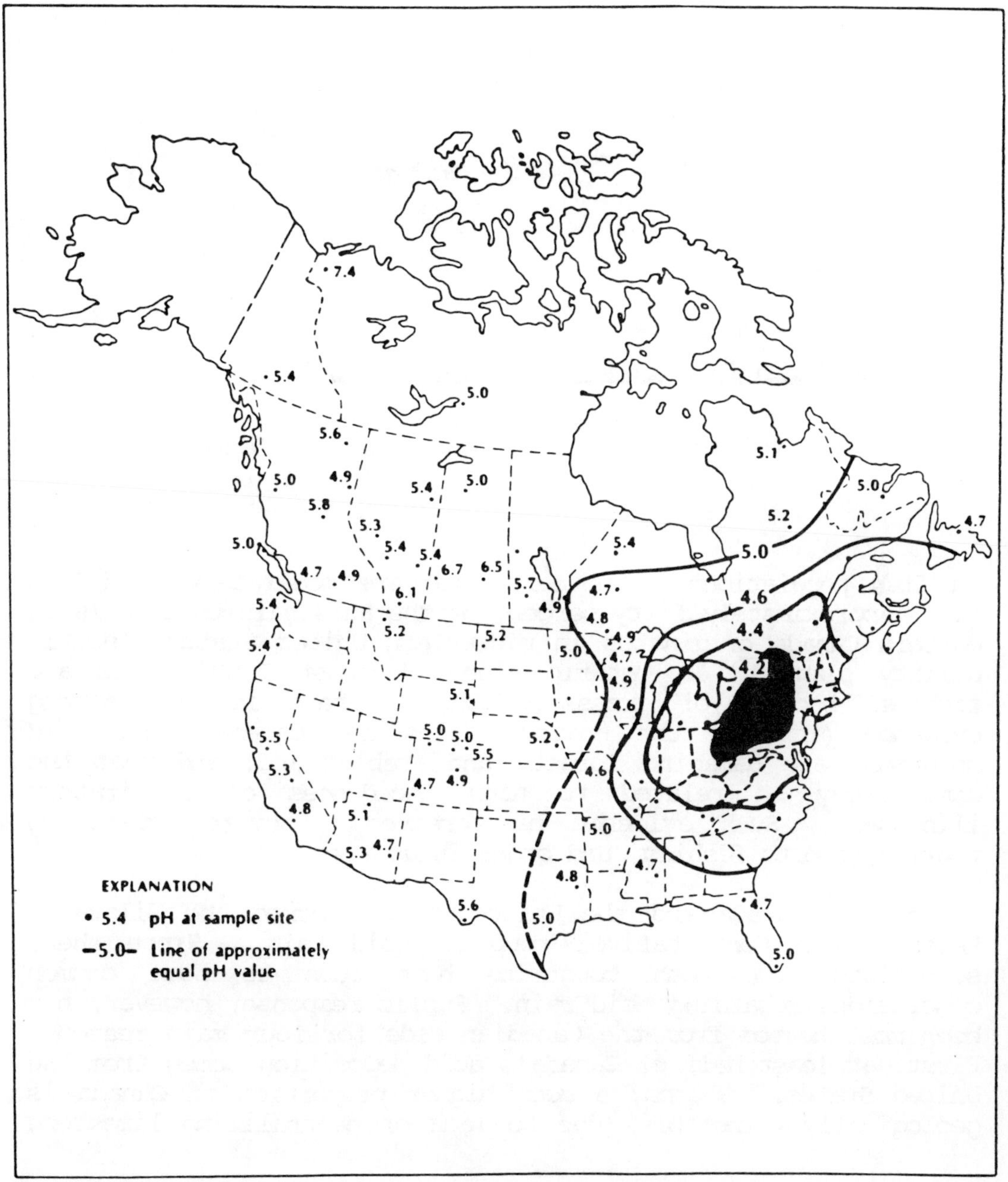

Figure 6-2. Typical Northeastern U.S. acid rain components.

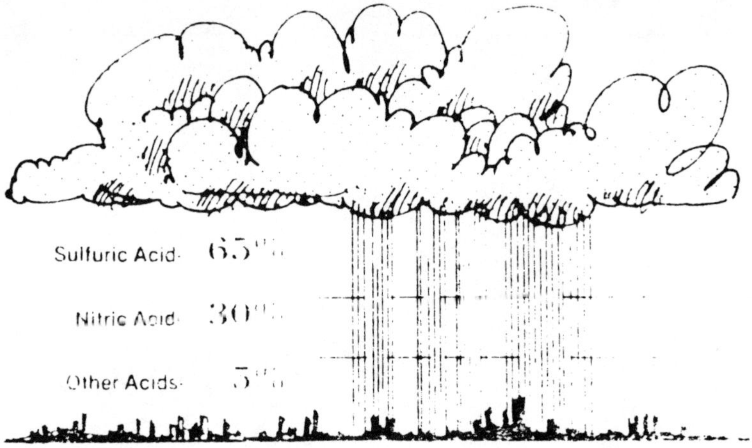

TYPICAL NORTHEASTERN U.S. ACID RAIN COMPONENTS

in fish populations. Another 3,000 are moderately acidified [4]. Approximately fifty percent of the acid rain that falls in eastern Canada originates in midwestern United States electric utility plants. In Ontario alone at least 1,600 lakes are fishless because of excess acidity. There is also strong evidence (in both the United States and Canada) that acid solutions are damaging forests and arable land, and that the damage may be related to high incidences of respiratory illnesses. Such evidence has created a severe impact on tourism, sports fishing, and other industries.

In 1978 Canada and the United States formed the Bilateral Research and Consultative Group on Acid Rain. Since then, scientists from both countries have come to some common conclusions regarding acid rain. Public response, however, has been most heated from the Canadian side for four main reasons. First, at least half of Canada's acid deposition comes from the United States. Second, a much higher proportion of Canada is geologically vulnerable (due to lack of neutralizing limestone exposures) to the impacts of acid deposition. Third, Canada is perhaps more dependent on its forest products industry, and acid deposition is having a negative impact on growth of Canada's commercially valuable softwood timber. The fourth factor is the disparity in concern expressed by people of Canada and the United States. Canadians generally feel Americans know little about the acid deposition problem and care even less. An added variable is the poor economic health of much of the region in the United States where emissions are generated -- the Great Lakes and Ohio River Valley -- making the cost of abatement difficult [3].

THE EUROPEAN SITUATION

Many European nations -- also recognizing that acid rain does not respect national boundaries -- have stepped up efforts to coordinate their clean air campaigns. Twenty nations have set targets for reducing their emissions of sulfur dioxide. The West Germans, for example, were galvanized into action by the impact of acid rain on the Black Forest. The West German government in 1983 set up environmental standards to cut domestic production of sulfur dioxide 65 percent by 1995, even though half the acid rain that falls in West Germany blows in from other countries. In addition, the United Kingdom, in response to complaints from Scandinavian countries downwind of British power plants, is aiming for a 25 percent reduction.

Buildings and monuments in Europe are also affected by acid rain. While stone structures are damaged, acid also weakens materials used in buildings and monuments, particularly iron, steel, zinc, and paint. The lifetimes of such structures are shortened appreciably.

Although emissions from natural and anthropogenic sources are of the same order of magnitude on a global scale, in Europe the number of anthropogenic sources greatly surpasses the number of natural ones. Between 1950 and 1970 European emissions of sulfur dioxide approximately doubled and then leveled off. During recent years sulfur dioxide emissions have decreased in many European countries, particularly in Western Europe. The present level is about 50 million metric tons (55 million U.S.tons) of sulfur dioxide per year.

Sulfur dioxide emissions are not evenly distributed over Europe (Table 6-1). Analysis of the distribution of emissions in 1980 reveals that a substantial portion of the sulfur dioxide is emitted in a highly industrialized area extending from southern parts of the United Kingdom over West and East Germany, and Czechoslovakia to Poland [2].

Because of the trend toward more efficient combustion processes, European nitrogen oxide emissions have increased relatively more than have sulfur dioxide emissions over the past decades. At present the total European emission is about 18 million metric tons (20 million U.S. tons) per year.

The spatial distribution of nitrogen oxide emissions is to some extent similar to that of sulfur dioxide emissions, reflecting the fact that, by and large, industrial activity, population density and road traffic are closely related. However, nitrogen oxide emissions are relatively higher in Western Europe than they are in Eastern Europe, largely because of the greater number of cars in Western Europe [2].

There are large differences among European countries in the importance of foreign sources, indigenous sources, and undecided background sources. In the United Kingdom, for example, about 80 percent of the sulfur deposition originates from emissions there, while foreign sources and ambiguous background sources contribute about 20 percent. For Norway the situation is very different. Norwegian sources contribute only 10 percent of the total deposition there, foreign sources 40 percent, while 50 percent is of undetermined origin.

The Swedish Society for the Conservation of Nature found that less than half of the sulfuric acid falling on Scandinavia emanates from Scandinavian chimneys. The rest they describe as being carried in on westerly and southerly winds from the United Kingdom and the mainland of Europe. In addition, the Swedish Ministry of Agriculture noted in 1983 that over 18,000 lakes

Table 6-1. Deposition from Foreign Sources of Sulfur Dioxide

(percent of total deposition)

Albania	57
Austria	73
Belgium	45
Bulgaria	29
Czechoslovakia	43
Denmark	47
Finland	34
France	32
Germany, East	31
Germany, West	46
Greece	33
Hungary	41
Iceland	5
Ireland	23
Italy	16
Luxembourg	75
Netherlands	65
Norway	40
Poland	42
Portugal	36
Romania	75
Spain	13
Sweden	44
Switzerland	73
Turkey	42
USSR	21
United Kingdom	10
Yugoslavia	53

SOURCE: [2]

were acidified (about 4,000 heavily) with severe ecological implications. Where waters are acidified, whatever the source, the content of highly toxic aluminum becomes enhanced. Species diversity decreases markedly, fish frequently become injured and mortality often occurs. In addition, a variety of other metal ions including zinc, lead, and cadmium become more readily soluble and consequently available to the fauna and flora of such affected ecosystems [5].

POLICY RESPONSES

Monitoring networks have aided in the improvement of the acid rain situation in Europe. The Monitoring and Evaluation of the Long-Range Transmission of Air Pollutants in Europe (EMEP), established in 1977 by the ECE (United Nations Economic Commission for Europe), now consists of a network of 92 monitoring stations in 24 countries. Results in Table 6-2 reflect the effectiveness of this pollution abatement policy.

PHYSICAL TECHNIQUES TO CONTROL ACID RAIN

A variety of physical techniques is available for reducing sulfur dioxide emissions from power plants, heavy industry and vehicles. These techniques fall into three categories: (1) precombustion, where before the fuel is burned, sulfur is removed through such techniques as coal cleaning, coal gasification and desulfurization of liquid fuels; (2) combustion, where sulfur removal occurs during the combustion phase in processes such as fluidized-bed combustion; and (3) postcombustion, where, after combustion, emissions are reduced by stack or flue gas desulfurization, or by scrubbers.

Nitrogen oxide controls for stationary sources involve modifying furnace and burner designs in hopes of reducing combustion temperatures. The control techniques employed are staging combustion, controlling air, injecting water during

Table 6-2. Total Sulfur Dioxide Emissions in Europe and North America in metric tons (millions) per year. (One metric ton equals about 1.1 U.S. tons).

	1980	1983	1985
Eastern Europe	23.5	----	17.8
Western Europe	27.8	----	26.9
North America	27.8	24.5	----

combustion, recirculating the gases, and altering the designs of the firing chamber. Nitrogen oxide production from mobile sources is reduced through lowering combustion temperatures in the engine and by catalytic removal from the exhaust. These nitrogen oxide control techniques are used to convert oxidized forms of nitrogen to molecular nitrogen, or to prevent the oxidation of nitrogen altogether [1].

Liming of lakes offers another solution to correct pH levels significantly lowered by acid rain. Liming increases the pH value of water and soils, thus reducing the acid level and making possible the restocking of fish and the maintaining of the natural ecosystem. In Sweden a national liming program has been implemented. Approximately 3,000 lakes and 100 streams have been treated in hopes of restoring the natural equilibrium to these water [8].

FUTURE DIRECTIONS

We have examined the extent of lakes acidified by acid rain, recognizing that such lakes no longer support fish populations and other aquatic life. We have also found that the numerous impacts from acid precipitation have led to a growing interest in pollution abatement programs. Emission data throughout the world are generally improving, mostly because of the growing awareness of transboundary air pollution problems and because of increasing cooperation among countries. Nevertheless, without a firm commitment both to developing improved pollution control methods and to applying these methods vigorously, the problem of the past could be a prelude to an acid rain future.

SUGGESTIONS FOR FURTHER WORK

1. Although acid precipitation is a serious environmental issue in North America and Europe, acid solutions create environmental degradation in other world regions. Investigate the impact of acid rain on countries outside North America and Europe.
2. The sources and impacts of acid rain vary spatially from the western to eastern United States. Compare and contrast different sections of the country in this regard.

SELECTED SOURCES FOR FURTHER READING

Southwick, C. (ed.). Global Ecology. Sunderland, Massachusetts: Sinauer Associates, 1983.

Cutter, S., H. Renwick, and W. Renwick. A Geographic Persepctive on Natural Resource Use, Totowa, New Jersey: Rowman and Allanheld, 1985.

REFERENCES

1. Carrol, J.E.. Acid Rain: An Issue in Canadian-American Relations, Durham, N.H.: Canadian-American Committee, University of New Hampshire, 1982.
2. Dorland, H.. "Monitoring European Transboundary Air Pollution," Environment, Vol. 29, No. 10, (1987), pp. 10-15 and 27-28.
3. Eney, A. and Petzold, D.. "The Problem of Acid Rain: An Overview," The Environmentalist, Vol. 7, No. 2, (1987), pp. 95-103.
4. Miller, G.T.. Living in the Environment, Belmont, California: Wadsworth Publishing Company, 1988.

5. Pyatt, F.B.. "Acid Rain in Scandinavia - Some Current Data," The Environmentalist, Vol. 7, No. 3, (1987), pp. 197-200.
6. Sand, P.H.. "International Policy Responses: Air Pollution in Europe, Environment, Vol. 29, No. 10, (1987), pp. 16-20 and 28-29.
7. Stopp, G.H., Jr.. Acid Rain, Metuchen, N.J.: The Scarecrow Press, Inc., 1985.
8. Thornelof, E. and O. Lessmark. "Living in Sweden," Water, Air and Soil Pollution, Vol. 31, (1986), pp. 809-814.
9. U.S. Environmental Protection Agency. Acid Rain, Washington: Office of Research and Development, 1980.

Chapter 7

Climatic Change

The Context

Everyone accepts the fact that the weather changes from summer to winter. However, the idea that the relative severity of winter cold or summer heat might be changing is less easily accepted. For example, if a location experiences a particularly dry or cold season, it is considered an unusual event, and most people assume that eventually wetter or warmer conditions will return. However, it is possible that the long-term conditions are actually changing, and pre-existing conditions may never return.

If such long-term changes were to occur, they would result in dire consequences for many people. Many of the basic needs of life, ranging from water requirements to agricultural production, would be modified and result in widespread hardship. Even though a change in climate might not produce a new ice age, it could threaten the livelihood of people in many parts of the world.

There is no doubt that such a variation could occur. We currently live in a post-glacial age, and there is no guarantee that the present climate is not changing. In fact, it is much more reasonable to assume that our climate represents a temporary phase and, in time, will become different. To examine such an idea, this chapter looks first at how climate has varied over time. To illustrate that fairly substantial changes can occur in relatively short time spans (in comparison to changes in geologic time) climate variation over the last 1000 years will be examined. The second part of the chapter examines the prospects for climates of the future.

CLIMATIC CHANGE OVER TIME

The three graphs in Figure 7-1 provide the essential details of climatic change over millions of years. Figure 7-1a places the most recent ice age in the perspective of geologic time. Here we see that for most of the eons of geologic time, average world temperatures have been appreciably higher than they are today. The Pleistocene ice ages and the period since ice retreated to its present polar position represent cooler climates than those that have existed over much of the earth's history. As shown in Figure 7-1b, the most recent ice age did not merely consist of a single ice advance followed by ice retreat; rather, it comprised a sequence of glacial times alternating with warmer, interglacial times, with the last retreat occurring about 10,000 years ago.

The last graph (Figure 7-1c) shows the variations in

Figure 7-1. Three graphs providing an overview of climatic changes over the last 600 million years. (a) Throughout much of geologic time, temperatures have been appreciably higher than they are today. (b) During the Pleistocene period, the most recent ice age, temperatures fluctuated to give glacial and interglacial conditions. (c) A representation of climatic changes over the last 10,000 years. The Climatic Optimum represents the warmest time since the retreat of the ice sheets.

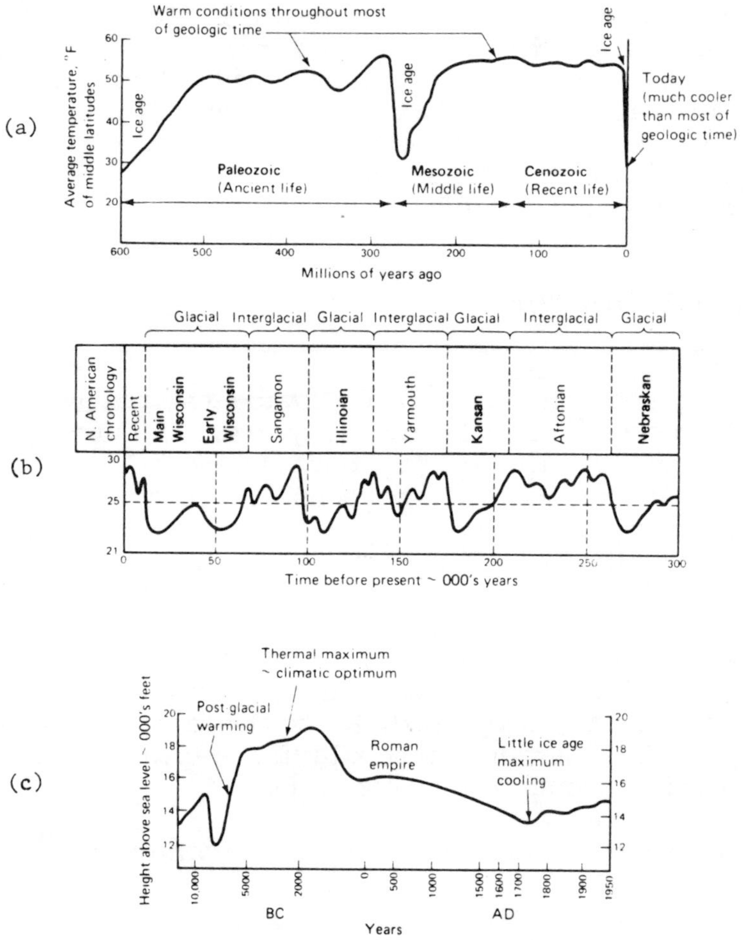

climate since the end of the Pleistocene Ice Age. It illustrates that there have been marked changes over the last 10,000 years. In this chapter we concentrate upon the events of the last 1000 years.

Before we examine the climate that has existed during the last millennium, we need to acknowledge an obvious question. Given the facts that until very recently (in geologic time) humans were not around to observe the weather and that some basic weather instruments were not even invented until the 17th century, how can former climates be determined?

Climatic reconstruction is, in many ways, a fascinating jig-saw puzzle to which some parts are still missing. Parts of the puzzle come from scientists of quite different disciplines, ranging from anthropologists through geologists to zoologists. The key to interpreting much of the past is through the use of proxy data; this involves the accumulation of information concerning, for example, fossils or types of rock accumulation, and inferring the climate that existed when the animals lived or the rocks were deposited. In recent years the interpretation has been greatly aided by the use of modern techniques including isotopes, atoms of the same elements that vary in their atomic weights.

An example of the use of isotopes is found in research using the ice of the huge ice caps of Greenland and Antarctica. The ice that forms layer upon layer, year after year, represents a historical record extending over thousands of years. By drilling through the ice using a special drill, a core is obtained. This core represents a column of the ice, and each of the layers in it may be analyzed. The water that comprises the ice contains oxygen, and oxygen has atoms that vary in their atomic weight from the normal atom; that is, there are isotopes of oxygen. The ratio of one oxygen isotope (oxygen 18) to another (oxygen 16) that is found in the ice depends upon the temperatures that prevailed when the ice formed. Thus, by analyzing the ratio of the isotopes in the ice, it is possible to estimate the environmental temperature at its time of formation, and an actual temperature can be derived.

Some of the many ways in which climate of the past can be determined using proxy methods are shown in Table 7-1. By using selected methods shown in the listing, it is possible to examine in some detail the climate of the last 1000 years.

The Last 1000 Years

As Figure 7-1c shows, subsequent to the retreat of the last ice advance associated with the Pleistocene period, temperatures warmed until, by about 2000 B.C., they attained a peak. This remarkable warming trend has been termed the Climatic Optimum. Since that time, global temperatures have not attained such warmth although during the last 1000 years a second time of relative warmth did occur.

The discussion of the last 1000 years can conveniently begin about 950 A.D.. From that time until about 1250, climate was mild over much of the middle latitudes to the extent that it has been termed the Little Climatic Optimum. To recreate the climates that existed since this Little Climatic Optimum, proxy data are derived from what are termed standard time series. These are chronologies built by using different interpretive methods to create a continuous record of climate at a location and, thereafter, to combine the chronologies to create a picture of the historical climatology. Only three chronologies will be used here to demonstrate the pattern of climate that existed.

The first graph in Figure 7-2 is based upon oxygen isotopes taken from ice cores in central Greenland, using the method already outlined. The second is derived from the North Atlantic island of Iceland where a fairly complete record has been derived from the sources listed below:

1864 to present	Actual meteorological records
1781 to 1845	A reconstruction of weather conditions obtained from the relative severity and frequency of drift ice in the vicinity of Iceland

TABLE 7-1

Sources of Proxy Data for Climatic Reconstruction

ICE (Glaciers and Ice Caps)
- Oxygen isotopes
- Physical properties of ice
- Trace elements in ice

GEOLOGIC EVIDENCE OF CLIMATIC CHANGE

Marine deposits
- Accumulation rates
- Mineral composition
- Geochemistry
- Oxygen isotopes
- Flora and fauna distribution
- Fauna and flora (morphological) variations

Land deposits
- Glacial erosion and deposition
- Periglacial features
- Wind deposits (loess and sand dunes)
- Ancient shore lines
- Lake deposits
- Relict soils
- Cave deposits

BIOLOGIC EVIDENCE

- Tree rings
- Pollen analysis
- Plant distribution
- Insect type and abundance
- Flora and fauna population distribution

HISTORICAL EVIDENCE

- Written records
- Phenological (crop dates) events
- Observation of frequency of events

Figure 7-2. A comparison of conditions, using three sets of proxy data. The graph on the left represents isotopic temperatures derived from Greenland ice cover; that in the center is a reconstruction of Icelandic conditions; the right hand graph is derived from reconstruction of records in Central England. Shaded parts of each graph represent warmer times.

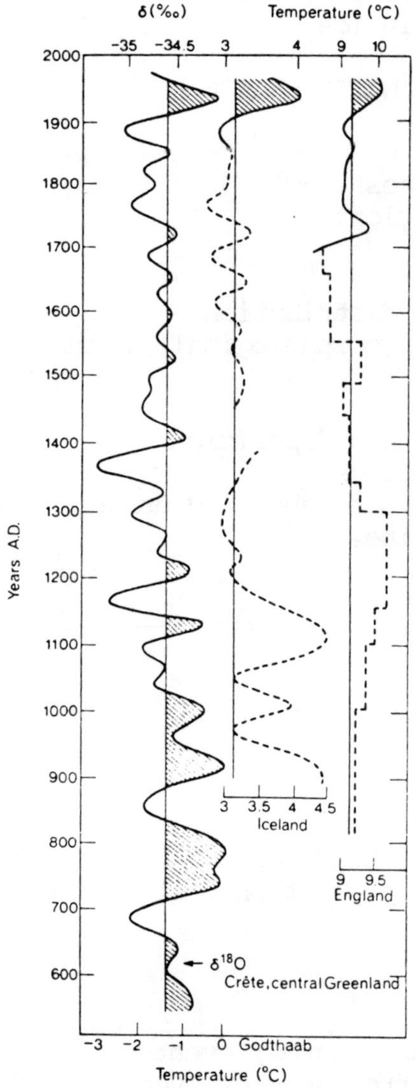

1591-1780	Historical records combined with incomplete drift ice data
900-1590	Information from Icelandic sagas indicating times of severe weather and related famines.

The third sequence shown in Figure 7-2 is the estimated course of temperature for central England since 800 A.D.. These data are obtained through examining actual data from the time when instruments were first available and using these to derive numerical indices expressing, in relative terms, reports of mild or cold, wet or dry, periods of earlier times. It is lengthy, painstaking work. The results are then examined in terms of decades or longer. Fifty-year blocks are considered fairly reliable back to 1100 A.D., but prior to that, 200-year means are used.

The three curves, while differing in detail, do show a similarity in the occurrence of cold and warm times. It is clear that up to about 1250 (later in the Central England record) warmer times prevailed, but this then gave way, with but a few exceptions, to a lengthy period of cooler times. In fact, it was not until about the late 19th century that the temperatures started to show any distinctive warming trend. These observations permit the period to be classed into four time periods:the Little Climatic Optimum (950-1250 A.D.); a period of change with deteriorating conditions (1200-1450 A.D.); the Little Ice Age (1450-1850 A.D.); and the warming associated with the Instrumental Period (1850 A.D.-Present).

The Little Climatic Optimum (950-1250 A.D.)

One of the most remarkable tales of exploration and settlement that occurred during this relatively warm time in the northern hemisphere concerns the Viking, Eric the Red. Living in Iceland, which had been settled by Norsemen in the mid-ninth century, Eric was banished from Iceland for three years for killing other Norsemen in a feud. Having heard of a land to the west, he set sail and for the next three years reconnoitered the coast of what is now Greenland. As the world's largest island, Greenland lies mostly within the Arctic Circle with most of its 2.2 million square kilometers (840,000 square miles) covered by an ice cap more than 300 meters (1,000 feet) thick. Yet, in 985 A.D., Eric convinced almost 500 people to sail with him from Iceland to establish new settlements in Greenland. Two colonies were founded, one on the southeast and the other on the southwest part of Greenland.

A number of points emerge from this saga. First, it must be remembered that the ships in use in the 10th century were small, sail- and oar-powered, and not well equipped to battle the icy conditions such as have existed in recent years. Second, and most significantly, for colonies to survive they had to be self-sufficient and provide food and clothing for the settlers. Consequently, it must be assumed that the masses of floating ice that exist in the Greenland waters must have been less than today and that conditions in coastal Greenland must have permitted farming and animal husbandry. Eric the Red and the settlers were fortunate to be living in the benign times of the Little Climatic Optimum.

However, and as discussed later, climate deteriorated and by 1250 A.D. Greenland was practically isolated from outside contact, with extensive drift ice preventing the passage of ships. By 1500 A.D., the settlements had practically been forgotten, and in 1540 A.D. a voyager reported seeing signs of them, but no signs of habitation. It appears as if the settlers had perished. Whether their demise was due to deteriorating climatic conditions or to invasions of other people is not known, although a Danish archeological expedition to the sites in 1921 A.D. found evidence that the worsening climate must have played a role. Graves were found in iron-hard permafrost (permanently frozen ground) that must have formed since the time of burial. Tree roots entangled in the coffins indicated that the graves were originally in unfrozen ground and that the permafrost had moved progressively higher. That food supplies had become insufficient was shown by examination of skeletons; many were deformed, and evidence of rickets and malnutrition was clear. Such evidence leads to the conclusion that the climate became progressively harsher, leading to the isolation and extinction of the settlers.

The relative warmth of the Little Climatic Optimum is also seen in many facets of life in Europe, but it is in agriculture that some of the major differences are recorded. Vineyards and wine-producing areas were much more extensive than they are today. Even in England, which now lacks productive wineries, there were thriving vineyards producing good quality wine. That this was the case is indicated by the fact that the French attempted (by treaty) to close down the English vineyards in order to decrease competition. It has also been noted that agriculture was possible on hill slopes at a much higher altitude than today.

While the changes in agriculture were taking place, the people of Europe were making great strides in other endeavors. This was a period when many of the beautiful cathedrals of Europe were initiated and a time which some historians consider the first great awakening in European civilization. Perhaps the mild climate of the period contributed toward the events that occurred.

Although less contemporary evidence is available in North America, it appears that the climate followed a similar pattern to that in Western Europe. One particularly interesting study of this period is reported by Bryson and Murray [1] concerning a site of the Mill Creek Indian culture in Iowa. As part of the study they examined bones remaining from the animals that were hunted and used as food by the Indians. As shown in Figure 7-3, in 900 A.D. the main bones were from deer (almost 70%) and elk (about 25%). Over time, the percentage of bones from these animals decreased until, by 1200 A.D., bison formed some 65% with elk and deer comprising the rest. By 1400 A.D., the time at which all evidence of the habitation at this site had vanished, no evidence is found of the bones of elk.

Figure 7-3. Percentages of elk, deer, and bison bones found at Mill Creek site. Notice the changes in frequency of bone types associated with the period, 900 to 1400 A.D. [After 1].

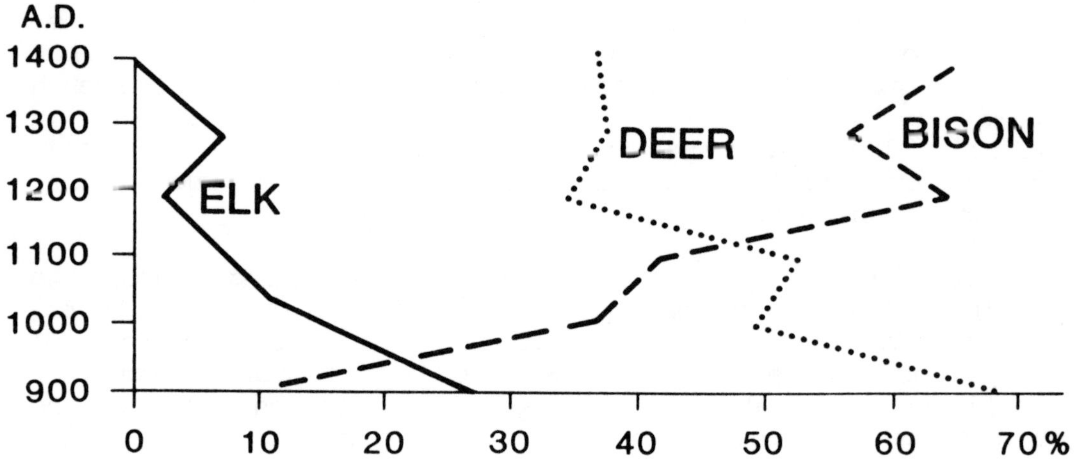

In their assessment of what happened, the researchers pointed out that deer and elk are forest animals, while bison are associated with grasslands and drier areas. It might be assumed that the meat diet of the Mill Creek people reflected an increasing dryness and that, over time, forests slowly gave way to grassland with increasing droughts. With even more drought conditions, animal numbers would decrease and hunting become increasingly difficult. As well as being hunters, the Indians were also corn farmers who used pottery to carry and store their grain. By counting the potsherds (pieces of broken pottery) at the site, it was found that the number decreased rapidly after about 1300 A.D., indicating both a decrease in the total population and an increase in adverse corn-growing conditions. In all, the community thrived during the time of the Little Climatic Optimum, but with increasingly difficult conditions were subsequently forced to migrate elsewhere. The Little Climatic Optimum ended and a climatic deterioration was underway.

The Late Middle Ages (1250-1450 A.D.): A Deteriorating Climate

During this period, climate became more erratic than in the preceding time, and sharp variations occurred from year to year and decade to decade. It has already been noted that the deteriorating conditions led to the end of the Viking Greenland settlements and caused the migration of Indians out of the Iowa region in North America. Merely by selecting a few other examples of the impacts of the adverse conditions, it becomes clear that the late Middle Ages differed significantly from the time preceding them.

Cooler temperatures led to a modified growing season for crops. In Iceland, grains were grown in the 10th Century. By the 16th Century they were totally abandoned. In England the shorter growing period and increased frost hazard of spring resulted in the failure of vineyards and wine production. Across much of Europe, cooler summer temperatures resulted in failure of grain harvests and privations for many people. The changing conditions led to the retreat of the limits of grain growth and the eventual abandonment of many once-thriving medieval villages and farms. Population declined in cooler areas as indicated by the example of Iceland. In 1095, the population of the island was 77,500; by 1311 this had declined to 72,000 and, in the even worse conditions that were to follow, the population in 1703 A.D. fell to 50,000.

Modified frequencies of climatic events resulted in widespread disasters. Exceptional floods and frequent storms led to flooding over wide areas; the Zuider Zee in Holland was formed during this time. In England some coastal towns, such as Dunwich in East Anglia, were inundated and abandoned. Wet conditions led to agricultural losses and possibly the spread of disease. One of the most horrifying diseases was ergotism or St Anthony's Fire, a disease that results from ergot blight of rye in damp harvests. About this H.H.Lamb [4] has written "Even a minute portion of the poisoned grains, baked in bread, would cause the disease. The course of the epidemics was such that the whole population of a village would suffer convulsions, hallucinations, gangrene rotting the extremities of the body, and death. In the chronic stages of the disease, the extremities developed first an icy feeling, then a burning sensation; the limbs then went dark as if burnt, shrivelled, and finally dropped off. Even domestic animals caught it and died."

This clearly was a period when the weather conditions changed significantly from the moderate times of the Little Climatic Optimum. But worse was yet to come.

The Little Ice Age (1450–1850 A.D.)

The precise meaning of the Little Ice Age has changed over time, but the term is now used exclusively to describe a cool period that began in the fifteenth century and lasted until the middle part of the nineteenth century. For our purposes we can use the dates 1450-1850 A.D..

There is both physical and historic evidence to support the cooling. Significant physical evidence comes, for example, from the advance and retreat of glaciers. During this period glaciers enlarged to the extent that some Alpine villages were overwhelmed by ice. The illustrations in Figure 7-4 show the village of Chamonix in the French Alps. The top figure is an engraving made in the 1700s showing how close the glacier was to the village. The lower is a photograph taken from the same vantage point in the early 19th century. The church and village appear similar, but the glacier has retreated far into the mountains. The data in Table 7-2 provide further evidence of the increase of ice during this period.

Figure 7-4. Changes in the mass of a mountain glacier near Chamonix, France. See text for details.

TABLE 7-2

Number of Days with Snow Cover
for Selected Locations in Switzerland

Winter	Lower Plateau	Upper Plateau
1769-70		126
1784-85	134	154
1788-89	112	
1962-63	59	86

The frequency of ice occurrence, in terms of winter freezing, also increased. The etching in Figure 7-5 shows an entire market on the frozen River Thames at London. Such a scene would be unlikely today, for complete freezing of this type does not occur. In fact, the relative severity of winters of the period can be assessed by the number of times that a given river freezes each century. In North America the freezing of large rivers was also observed. During the American Revolution, British soldiers were able to slide their guns across the frozen River Hudson from Manhattan to Staten Island.

It is from North America that there comes a well-known account of life during the last years of the Little Ice Age. This is a description of the year 1816, known as the "year without a summer." The year began with excessively low temperatures across the eastern part of the country and was followed by a spring that was cool, but not excessively so. In May, however, the temperatures plunged. In Indiana there was snow or sleet for 17 days, which killed off seedlings before they had a chance to grow. The cold weather continued into June when snow again fell, devastating any remaining crops. It is reported that no crops grew north of a line between the Potomac and Ohio rivers, and diminished yields were obtained south of that line. In the pioneer areas of Indiana and Illinois the lack of crops meant that hunting and fishing had to supply the food. Reports suggest that raccoons, groundhogs and the easily trapped passenger pigeon were major sources of food. The settlers also collected many edible wild plants which had proved hardier than cultivated crops.

Figure 7-5. Scene on the River Thames, at London in the 17th Century. The river was sufficiently frozen to permit temporary structures to be built.

Historic reports of the impacts of the Little Ice Age are extensive, as well shown in the accounts of Lamb [4] and Ladurie [3]. It has also been suggested that the relative severity of the period is reflected in the art of the time. In a study of paintings completed during the Little Ice Age, Hans Neuberger [6] has shown that in depicting atmospheric features, ranging from blueness of sky to cloudiness, artists provide a perception of a climate cooler and more stormy than that of today. The painting shown in Figure 7-6 by Pieter Brueghel, the Elder, shows ice and snow as the central theme in the Low Countries. It is probably typical of the more rigorous climate that existed at the time that it was painted.

The Instrumental Period (1850-Present)

Many basic meteorological instruments have been in use for several hundred years. The thermometer and the barometer were invented in the first half of the seventeenth century, while the rain gauge and wind vane may well have been in use prior to that time. However, there was no standardization of instruments or units used, and the observations were limited, to say the least. In fact, it is only from about 1850 that measurements began to resemble anything like a network that could be used for interpreting climate. The dynamic explanation of climate came much later when upper air measurements were initiated on a routine basis in the 1940s. Nowadays, satellites and automated stations provide a fairly comprehensive coverage of the world.

Using actual data, the study of climatic change over relatively short time periods can employ statistical methods to evaluate the nuances of climate. Using available statistics a number of authors [e.g. 4] have analyzed the trends that have occurred; results obtained for North America are shown in Figure 7-7. The pattern that emerges is an interesting one. From 1895 temperatures showed a warming trend, with the 1930s representing the time when most years were above the long-term normal. In the early 1940s a cooling trend began, with the period from about 1955 to 1979 showing a downward trend. Since this graph was constructed, values for the 1980s indicate that the cooling has ended and average temperatures have increased. In fact, the 1980s have contained the five warmest years of the last century.

These increases and decreases over the last century do not represent huge temperature changes, most being in the order of a fraction of a degree Celsius. But, as indicated in the following section, such changes are highly significant.

Figure 7-6. The Hunters in the Snow by Pieter Brueghel, the Elder.

CLIMATES OF THE FUTURE

It will have been noted that the discussion of climates of the past has concentrated upon changes in temperature. The reason for this is related to the fact that patterns of weather that occur at a location are related to air masses and their interaction, and these are contingent upon the nature of the upper air circulation that exists. As a guide to upper air circulation, jet stream location is used and this, ultimately, is determined by the equator to pole temperature gradient that occurs over the earth. The greater the difference between

Figure 7-7. Northern hemisphere average temperatures, 1895-1983. The dashed lines are merely suggestive of trends and do not represent precise statistical values.

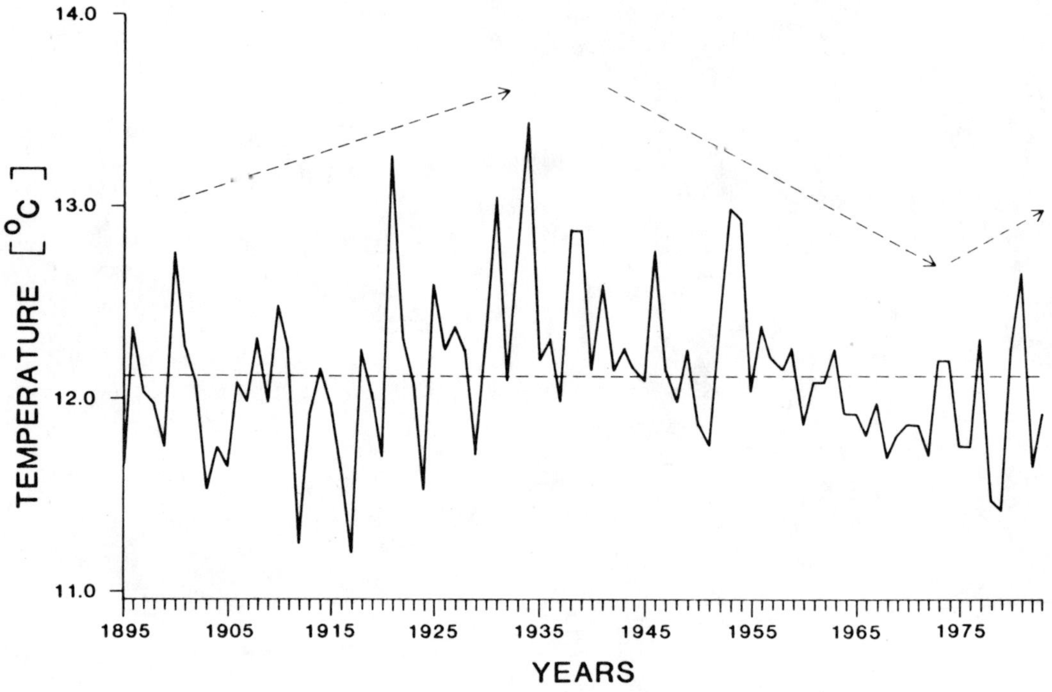

prevailing temperatures in the tropics and poles, the further south the upper air westerlies, as indicated by jet streams, will be. Small gradients result in the jet stream being located further poleward. Thus, the variations in temperature that occur influence climate by modifying the atmospheric circulation patterns.

Figure 7-8a illustrates this concept. The boundary between tropical circulation (the Hadley Regime) and middle latitude circulation (the Rossby Regime) is indicated by the subtropical jet stream. If the equator to pole temperature gradient varies,

Figure 7-8. Changes in the equator-pole temperature gradient result in modified circulation: (a) location of jet streams, indicating the equatorward limit of the Rossby regime, as a function of warming and cooling trends; (b) some potential impacts of a modified temperature gradient and resulting circulation changes.

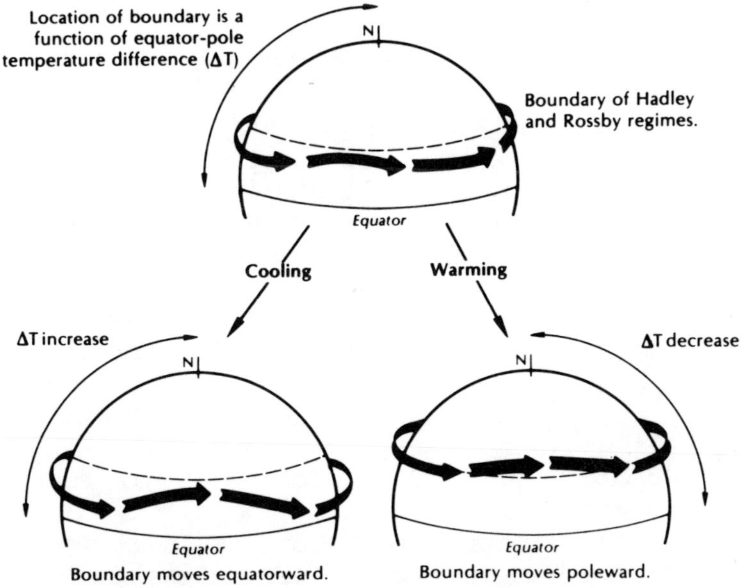

(a)

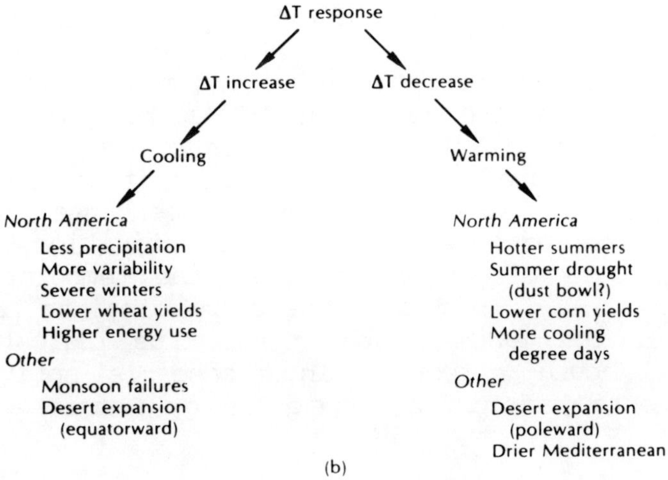

(b)

then the position of the jet changes; this will lead to climatic modification over wide areas as indicated by the examples listed in Figure 7-8b.

It remains then, to determine whether the prevailing temperature is going to rise or decline in the years ahead. To obtain a potential answer to this requires much theoretical work and the use of data that have been collected over time. Of increasing importance in the evaluation is the use of models, especially those that use high speed computers.

Various types of models are employed. Analog models use a set of conditions that are analogous to those of the real world, a somewhat difficult thing to replicate in all its detail. As a result, digital models are used more frequently. Some of these utilize fundamental equations that describe the behavior of gases and forces involved in the atmosphere. Again, the complexity of the atmosphere means that even the largest computer can barely handle all the variables for all points on earth, and most use a limited number of points spaced over a grid.

The goal of these and other models is to reproduce the dynamics of the general circulation of the earth. Once, and if, this is done, then climatic prediction could become much more exact. Until that time, however, a useful approach to dealing with future climates is to establish scenarios. A scenario attempts to answer the question "What would happen if . . .?", with the scenario establishing the "if" part of the question.

A scenario that is currently receiving much attention is that which depends upon the addition of carbon dioxide to the atmosphere. Carbon dioxide forms an important, although tiny, fraction of the natural atmosphere in that it inhibits the loss of earth heat to space through the greenhouse effect. By addition of carbon dioxide through human activities, enhancement of the greenhouse effect would raise the temperature of the earth's atmosphere. In fact, some climatologists believe that the warmth of the 1980s is the first true indication of the enhanced greenhouse effect. This case is aptly described through comments made to a Senate Energy Committee, by James

Hansen of the NASA Goddard Institute of Space Studies. He indicated that according to records going back about a century, global temperatures have increased by 0.6-0.7°C (about 1.1°F), and the five warmest years in the last century have occurred in the 1980s [5]. Much of this is attributed by Hansen to the greenhouse warming effect.

The Current View

At the present time many researchers are attempting to provide a method whereby climate in the coming century might be forecast. It is a very difficult task, and differences in opinion certainly exist. However, a consensus appears to hold the following viewpoints:

> World climates have changed; they are changing now and will change in the future.
>
> The cooling trend that began in 1950 appears to be over, and it is probable that by the year 2000 a slight warming will be distinguishable from normal climatic variations that occur.
>
> If unusual occurrences (e.g. widespread volcanic activity) that could induce cooling do not occur, then the mean temperature of the earth by 2050 A.D. will be greater than at any other time over the last 1000 years.
>
> The changes will not be evenly distributed over the earth, and regional climatic changes will differ. It is expected that polar regions will have changes two or three times the global average, and this will influence circulation patterns and precipitation distribution.

Given such a forecast, the impacts of climatic change will be widespread.

The Potential Impacts

The climate that occurs at any location influences the environment that exists and the activities of the inhabitants who live there. By selecting but a few examples, we can get some indication of the possible impacts of climatic change.

Energy provides the basis for the economy of any society. In recent years much has been written about the problems of future energy resources, particularly those that are considered non-renewable resources (see Chapters 14,16, and 18). Problems of energy needs will be heightened by climatic change. The potential impacts can be demonstrated using energy demand by residential and commercial properties as an example. Currently, the United States uses almost 20% of its total energy demand for space heating and cooling. The amount used depends largely upon the outdoor temperatures and the type of building being heated or cooled, with the former being most significant over the long term. It might be assumed that a global warming would result in a decrease of energy demands. That such is not the case is seen in summer "brown-outs" that occur in cities during hot summer periods. Air conditioning is a high energy user, and increased summer heat will significantly increase energy demands.

Food Production depends upon an agricultural system that is tuned to the climate that prevails. Consider, for example, the production of corn in the Corn Belt states. The long-term mean summer temperature of the area is 22.2°C (72°F) in June, 24.4°C (76°F) in July, and 23.3°C (74°F) in August. Research has shown that highest yields are obtained when the mean is 22.2°C (72°F) with sufficient rainfall. Thus, a slight cooling of summer temperature would not drastically reduce yields, whereas an increase most certainly would. Given the warming prognostication, it might be anticipated that the yields from the highly productive Corn Belt will diminish over the years ahead.

As noted, however, changing temperature regimes result in modified circulation patterns. Given this, it must be anticipated that many areas currently in marginally moist areas would experience a change. The direction of the change (to more or less rainfall) would depend upon the nature of the climatic change. With the anticipated warming, the climate of California would become drier than it already is, while increasing summer drought would again raise the specter of dust bowl conditions in the central United States. Deserts would expand poleward, and overall summers would be warmer and drier. The results upon the agriculture of these locations would be disastrous.

Global Ecology of necessity would undergo changes. Unlike agricultural systems, in which a limited number of plant species are managed by people, natural ecosystems consist of a diversity of animals and plants that exist in balance with the prevailing environmental conditions. When these conditions change, the nature of the ecosystem is altered.

Evidence shows that massive changes in world ecosystems have occurred since the retreat of the ice cover of the Pleistocene period. At the height of the ice ages, tundra and subarctic climates extended across much of North America, presenting a very different environment from the forests and grasslands of today. Such changes reflect a long-term response of vegetation cover to changing climates and, if "abnormal conditions" were to come about, the current vegetation would also change.

If, as predicted, climates were to become warmer, then the current limits of ecosystems would shift poleward. While some subarctic climates might become less severe, the poleward expansion of deserts would impinge upon more densely populated parts of the United States. Such changes would result in modified vegetation and animal distributions. The extensive changes that could occur certainly permit the use of the scenario approach to what the future will be like.

These three examples provide but some of the potential impacts of a changing climate. Other impacts range from that upon the fishing industry to problems associated with health and nutrition. It is thus fortunate that the US Congress was farsighted enough to provide funds for research into future climate (National Climate Program Act, 1979). It is to be hoped that research generated by the Act will help diminish the potential hardships that could be created by changing climates.

CONCLUSION

Climates have changed over time, and there is no reason to believe that the current conditions will remain the same. The experiences of the past 1000 years clearly show that it does not take the formation of ice caps, or their melting, to have major impacts upon humans and their activities. When looking to the future, one can see evidence pointing to the probable warming of climates. While this may seem, at first, to be an asset, there are many problems that will need to be faced. Perhaps the most significant of these is the way in which a changed climate will influence agriculture, for the food to support large populations is of critical importance.

The study of climatic change is no longer an academic exercise, it is an area of study that has profound impacts upon the future. While not all answers are currently available, great strides have been made in the efforts to both understand and predict climate.

SUGGESTIONS FOR FURTHER WORK

1. Obtain climatic data for your area from the National Climatic Center or another readily available source. Plot the data and determine climatic trends over time. Use graphical analytic techniques (as given for example (below) in Oliver, 1981) to identify the trends.

2. Assume that world temperatures will increase significantly by the year 2050. Given the change, establish scenarios for some of the potential impacts.

SELECTED SOURCES FOR FURTHER READING

Bradley, R.S. Quaternary Paleoclimatology, Winchester, Massachusetts: Allen and Unwin, 1985.

Critchfield, H.J. "Climatic Data: Sources", in J.E.Oliver and R.W. Fairbridge. The Encyclopedia of Climatology, New York: Van Nostrand Reinhold, 1987.

Kates, R.W., J.H. Ausubal and M. Berberian. Climate Impact Assessment, SCOPE Publication 27, New York: John Wiley, 1985.

Oliver, J.E. Climatology: Selected Applications, New York: Halstead Press, 1981.

Oliver, J.E. and R.W. Fairbridge, (eds.). The Encyclopedia of Climatology, New York: Van Nostrand Reinhold, 1987.

Wilhite, D.A., W.E. Easterling and D.A. Wood. Planning for Drought: Toward a Reduction of Social Vulnerability, Boulder: Westview Press, 1987.

REFERENCES

1. Bryson, R.A. and T.J.Murray. Climates of Hunger: Mankind and the World's Changing Weather, Madison: University of Wisconsin Press, 1977.
2. Jones, P.D., et al. "Northern Hemisphere Surface Air Temperature Variations: 1851-1984," Journal of Climate and Applied Meteorology, Vol. 25 (1986), No.2, pp. 161-179.
3. Ladurie, LeRoy E. Times of Feast, Times of Famine: A History of Climate Since the Year 1000, (translated by B. Bray). London: Allen and Unwin, 1972.
4. Lamb, H.H. Climate, History and the Modern World, London: Methuen, 1982.
5. Monastersky, R. "Scientist Says Greenhouse Warming is Here," Science News, Vol. 134 (1988), No. 1.
6. Neuberger, H. "Art and Climate," in J.E. Oliver and R.W. Fairbridge, The Encyclopedia of Climatology, New York: Van Nostrand Reinhold, 1987.

Part III
The Hydrosphere

Chapter 8

Perspectives on Floods and Flooding

The Context

The study of water in all its forms is very important to geographers. From the moisture in the air that forms clouds to the erosion of the earth's surface by rivers, water plays a large role in physical geographic processes. However, it is also important to understand how water directly affects the lives of many people.

In the chapter we shall examine one aspect of water, namely floods, and consider some of the environmental and social problems that arise from flooding. The chapter is divided into three sections. First, we shall look at the frequency with which floods occur and the possibility of identifying floods according to their magnitude. The second section is devoted to an analysis of one way in which people have tried to modify river channels so that the risk of floods is lessened. Like any substantial program that modifies the natural environment, this method is not without problems. Finally, we shall examine the social aspects of flooding: the attitudes people living in flood-prone areas have toward floods. These three aspects of flooding clearly show that the understanding of a natural hazard concerns both its physical aspects and the way that people react to it.

THE MAGNITUDE OF FLOODS

A colleague of mine wanted to buy a house. After much searching he found what looked like the ideal home. It was set on a gentle slope up away from a small tributary of Indiana's Wabash River, located about 400 meters (a quarter of a mile) from the house. The small stream flowed in a channel below banks about one and a half meters (five feet) high. It looked like a pleasant stream, lazily flowing through wooded glades. The man was well informed and knew of the flood problems in the Wabash Valley. He asked the home owner if floods were a problem at the house location. The home owner replied: "I've lived here eleven years and water has never come out of the creek; there's no problem." So my colleague bought the house. Two years later, neighbors, the local fire brigade, and bands of other concerned citizens were frantically putting sandbags in front of the house, as water from the creek inched its way up the slope. Fortunately, no severe damage resulted. The man who sold the house was not dishonest. In the time that he had lived at that location there had been no problems. He and the house buyer had forgotten that floods recur at intervals.

Floods occur whenever a stream channel is unable to contain the volume of water being discharged along it. Most of the time the channel is large enough to contain the flow, but when intense precipitation or rapidly melting snow supplies vast quantities of water that cannot be contained in the channel, the river overflows its banks. The amount of water that overflows naturally depends upon the volume of water in the river, and during heavy storms, great volumes of water are available. Such storms may seldom occur with the result that high floods associated with them may never have been observed. On the other hand, lower amounts of rainfall may give "small" floods once every five years or so. In such a case the flood level is said to have a recurrence interval of five years. In the case of the flooded Wabash tributary, the original home owner had simply not lived at the location long enough to observe flood levels associated with, say, a twenty-year recurrence interval.

This recurrence interval is based upon the magnitude of floods that occur over a period of time. They are therefore referred to as a 10-year flood, a 50-year flood, or a 100-year flood, meaning, in each case, that floods of a given magnitude occur on the average once in 10 years, 50 years, or 100 years. It does not necessarily mean, however, that the floods occur exactly at 10-, 50-, or 100-year intervals. Perhaps a better way of looking at flood recurrence is to say that a 10-year flood has one chance in ten of occurring in any year.

The recurrence interval is easy to work out. The peak flow or flood of a river is recorded for a number of years. The flows are then listed by their magnitude or size, and each value is given a recurrence interval. This interval is provided by the formula (N+1)/M, where N is the number of years of record and M is the order of the flood's magnitude. Table 8-1 shows how this works [7]. In 1954, for instance, the peak flow of the Black Vermillion River in northeastern Kansas was the twenty-fifth highest recorded over the period of observation. The recurrence interval for this discharge, or flow, is:

$\frac{(N+1)}{M}$, where N is the length of record (33 years) and M is the rank of the 1954 flood (25).

$$\text{then } \frac{33+1}{25} = \frac{34}{25} = 1.36$$

Thus, the recurrence interval is 1.36. For the highest flood

Table 8-1. Peak Flow, Its Rank Order, and Recurrence Interval for Each Year of Record, 1954-86, Black Vermillion River near Frankfort, Kansas. (Note: cubic feet – 20.6415 = cubic meters)

Year	Peak Flow (cfs)	Rank Order	Recurrence Interval (yrs) $\frac{N+1}{M}$ $\frac{33+1}{M}$
1954	4,210	25	1.36
1955	2,840	28	1.21
1956	1,890	31	1.10
1957	1,400	33	1.03
1958	12,800	12	2.83
1959	38,300	1	34.00
1960	25,100	4	8.50
1961	4,760	24	1.42
1962	27,400	3	11.33
1963	2,870	27	1.26
1964	16,200	8	4.25
1965	11,200	14	2.43
1966	1,870	32	1.06
1967	16,600	6	5.67
1968	4,890	23	1.48
1969	7,000	19	1.79
1970	13,400	11	3.09
1971	2,240	30	1.13
1972	2,720	29	1.17
1973	12,400	13	2.62
1974	36,400	2	17.00
1975	6,320	20	1.70
1976	5,270	22	1.55
1977	16,600	6	5.67
1978	7,570	17	2.00
1979	8,560	16	2.13
1980	7,090	18	1.89
1981	3,640	26	1.31
1982	19,700	5	6.80
1983	15,400	9	3.78
1984	14,200	10	3.40
1985	6,190	21	1.62
1986	9,760	15	2.27

Sources:

1954-60--U.S. Geological Survey, Compilation of Records of Surface Waters of the United States, October, 1950 to September, 1960, Part 6B, (Water Supply Paper No. 1730), (Washington: U.S. Government Printing Office, 1964).

1961 --__________, Surface Water Supply of the United States, 1961-65, Part 6, (Water Supply Paper No. 1919), (Washington: U.S. Government Printing Office, 1969).

1962-63--__________, Surface Water Records of Kansas, (Topeka: U.S. Geological Survey, 1962, 1963).

1964-86--__________, Water Resources Data for Kansas, (Topeka: U.S. Geological Survey, 1965ff).

(rank 1, in 1959), the interval is 34/1=34 years. As with much hydrologic information, results are more clearly shown in graphs, and Figure 8-1 shows the Black Vermillion River data plotted on a flood frequency curve. The ordinate (X axis) is shown on a linear scale, the abscissa (Y axis) on a logarithmic. This semi-logarithmic form allows widely spread data, ranging from 1,000 to 40,000 cubic feet per second on the abscissa, to be included on a single graph.

The problems associated with analysis of flood frequency are evident in the formula used. All important is N, the length of record. Without a lengthy period of observation the recurrence interval does not mean very much, and even if a long record is available, it can ba misleading. Consider the example of the Connecticut River at Hartford [5]. Data collected between 1840 and 1936 show that the highest flood recorded was in 1854. These data represent a recurrence interval of almost 100 years (Figure 8-2a), but investigation of historic data showed that the 1854 flood was the greatest recorded since 1683. The flood in 1854 can therefore be interpreted as a 300-year flood, and the graph for the recurrence interval can be extended to show the level anticipated for a 300-year flood. Given such a time span, it might be assumed that flood levels higher than those recorded would be highly unlikely. Unfortunately, in 1936 a flood appreciably higher than that of 1854 occurred, and in 1938 another major flood also surpassed the 1854 event. As of 1976, the 1936 flood was still the event of record [8:13].

When these two Connecticut River floods were added to the records, the recurrence interval for given volumes of water had to be modified. Figure 8-2 shows the required modification. The new line is appreciably steeper, and the increase in the volume of water for the 300-year interval is 12 percent greater than was formerly thought. It is obviously risky to try to predict the possible discharge of a 500-year or even a 1,000-year flood. Yet floods of such magnitude could happen tomorrow!

Since many rivers are subject to flooding, and since records do not extend over long enough periods to estimate recurrence intervals greater then a hundred years or so, investigators have looked for other methods to deduce recurrence intervals. One such study was completed by Helley and Lamarche in 1968 [4: D34-D37]. They noted that in 1955 and 1964 unusually high floods on Blue Creek, a tributary of the Klamath River in northern California, and the usual method of calculating the flood recurrence interval failed to explain why the 1964 flood was the highest ever recorded. It seemed that the recurrence level for the river was longer than the predicted 115 years. Helley and Lamarche thought it might be as long as 200 years. To answer this intriguing question, the researchers turned to botanical and geomorphic evidence on Blue Creek.

On a high terrace adjacent to Blue Creek, erosion from the 1964 flood, which covered the terrace, revealed stumps of redwood trees. The tops of the stumps were all level, having been logged in 1958-59. At the time of the logging the trees were dead and limbless. Tree ring counts showed that the trees were 500 to 600 years old when they died. By examining the deposits on the terrace and by studying the structure of the tree roots, they found that the trees had been buried below 20 feet of sediment. To bury the stumps in this fashion required a flood of the magnitude of the 1965 flood. To determine when such a flood occurred they counted tree rings and dated some of them, using radiocarbon analysis. They established that the older trees began growth about 1000 A.D. and died about 1560.

Figure 8-1. Graphical Representation of Table 8-1. (For sources see Table 8-1.). Note: $\frac{\text{cubic feet}}{20.6415}$ = cubic meters

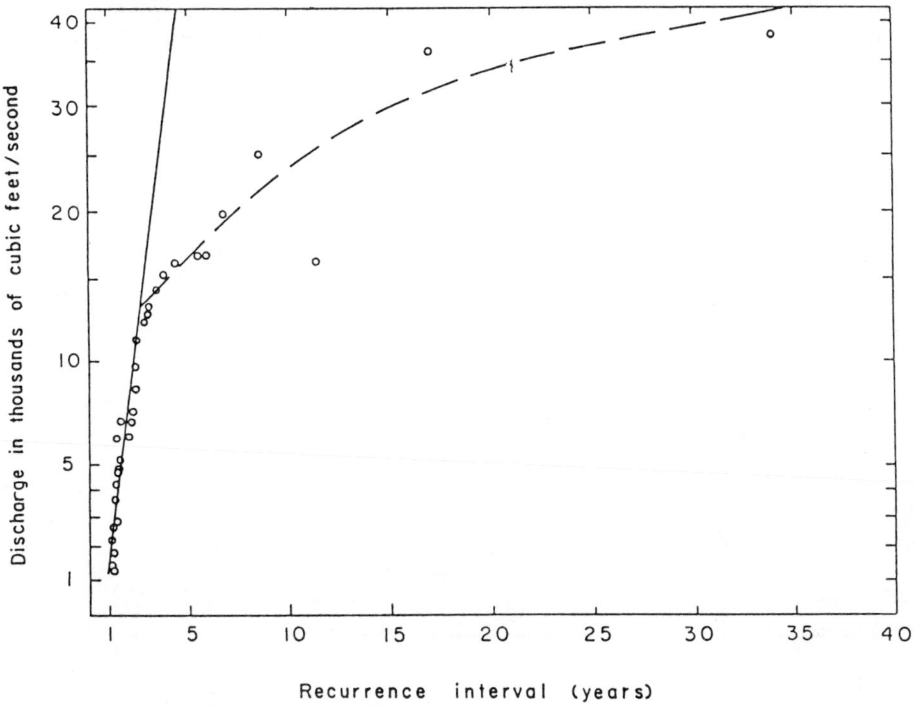

Since all the evidence pointed to their death by flooding, the devastating flood must have occurred 400 years ago, plus or minus 100 years. The results of the research show that the 1964 flood, using the older method of recurrence-level analysis, would have been expected to recur at an interval of 115 years. In fact, the flood was the greatest since the destruction of the redwood trees some 400 years ago. It was a 400-year flood.

Figure 8-2. a. Flood frequency graph for the Connecticut River prior to the 1936 flood. b. After the 1936 Flood the graphical representation of the long-term recurrence intervals changes appreciably. (From Floods, by William G. Hoyt and Walter B. Langbein (copyright 1955 by Princeton University Press) Fig. 19, p. 59 "Annual Floods of Connecticut River at Hartford, 1841-1950" in slightly modified form. Reprinted by permission of Princeton University Press.) Note: stage in feet x .3048 = stage in meters.

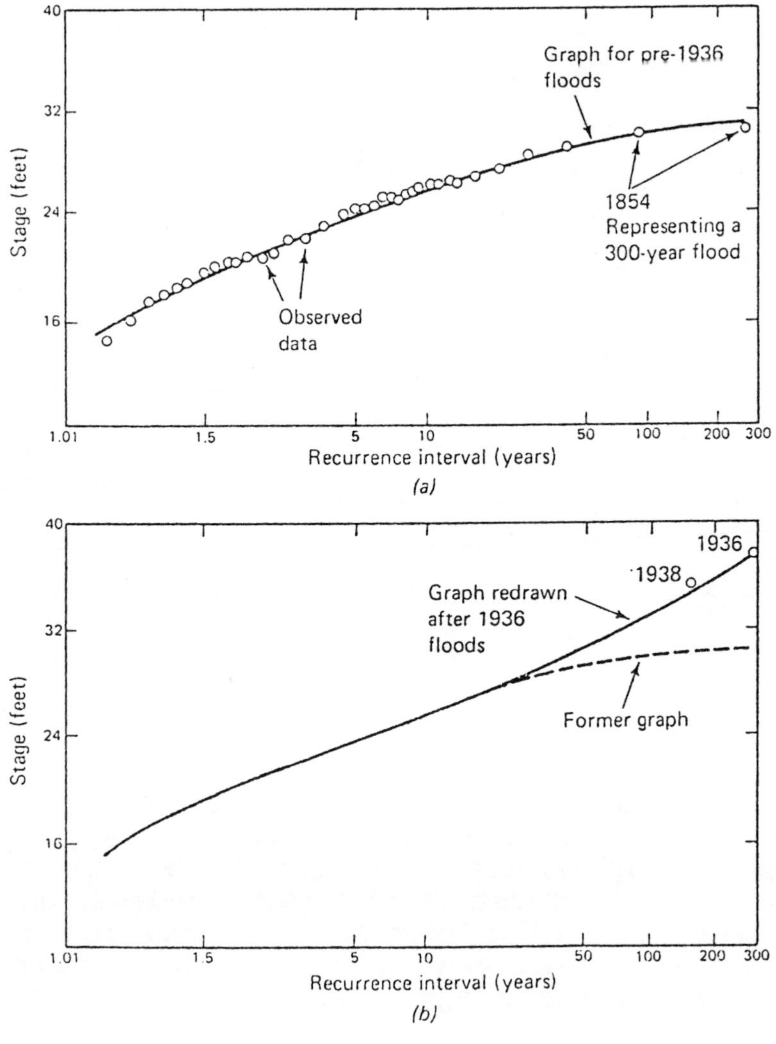

CHANNELIZATION

Rivers have played an important role in the history of human beings. Unfortunately, people have always used the easily flooded land on either side of rivers for settlements or farming. Later in this chapter we shall explore more fully the reasons why people live near rivers. At this point I wish to assess a method whereby floods can be minimized in such areas.

Many schemes to minimize floods have been initiated. Extensive levee or dike construction, dam building, and watershed management have all been used at various locations in the United States. The case we shall examine here concerns the role of straightening a river channel, channelization, and the controversy that the process has created. Channelization involves the straightening of channels so that they become wider and deeper than the natural course. The object is to shorten the distance traveled by floodwaters and hence to promote a rapid movement downstream. The movitation is economic, for channelization aims "to protect the land from floods, improve navigation, and to help private landowners drain tracts of marsh and swamp and the rich wet hardwood forests that thrived along the floodplains of the southeastern United States so that new land might be opened to cultivation" [3:890-894]. Although channelization has gone on for many years, it was not until the 1954 Act for Watershed Protection and Flood Prevention that the process really began.

There is little doubt that economic gains have been achieved by the program. Local flooding has been minimized, new land in the floodplain brought into cultivation, and swamps drained. The case against channelization is that, in the long run, the biological degradation, the shrinkage of wetlands, and the geomorphic response of the stream to its modification will outweigh all of the immediate economic gains. Two examples will illustrate the controversy.

Crow Creek was a scenic stream that flowed through the Appalachian Mountain valleys of Tennessee to northern Alabama. Its banks were lined with trees, and its waters provided excellent trout fishing. But the waters of Crow Creek emptied into the huge Guntersville Reservoir of the Tennessee Valley Authority (TVA). When heavy rains occurred, water from the reservoir backed up and inundated the stream's floodplain and the associated farm land. Residents of the ares most affected by flooding prevailed upon the Soil Conservation Service (SCS) to channelize the stream.

In drawing up the channelization plans, which took some five years (largely because of objections of environmentalists), many provisions were made to mitigate ecological disturbances. By usual standards the planning of Crow Creek Channelization was rigorous. Yet when biologists from the Philadelphia Academy of Natural Sciences visited the scene, they found:

> soft clay banks, stripped bare of vegetation, [that] crumbled into the muddied stream even as they watched. A long search could find no rooted plants in the streambed and no established populations of fish or other animals. A few aquatic creatures and rotting insect pupae were found stuck to the wet clay banks, now deeper and wider than ever before. Black flies had become the 'dominant organism,' but even they had evidently drifted from untouched reaches far upstream [3: 890-894].

All of the provisions to mitigate environmental damage had failed (Figure 8-3).

It seems, too, that the environmental damage might well outweigh any economic benefits derived from the project. Based upon a life expectancy for the channelization of fifty years, the cost of the project ($1.13 million in 1972) breaks down to an expenditure of about $21 per hectare ($8.50 per acre) for the area now protected from flooding. Income derived from the use of that land is estimated at $27 per hectare ($11.00 per acre) per year. However, the expenditure estimate does not include a financial statement for the ecological damage sustained.

A second example of channelization shows how this method of flood control results in modification of the way in which streams carve their channels. The Blackwater River is a

Figure 8-3. Crow Creek before and after its channelization. (Photos courtesy of Tennessee Wildlife Resources Agency).

tributary of the Missouri, the confluence occurring just west of Boonville, Missouri. Before 1910, when the river was channelized, the stream meandered across Johnson County at an average channel gradient of 1.67 meters per kilometer (about nine feet per mile). The hydrologic dimensions of the old river are schematically shown in Figure 8-4a. With the construction of a new channel, the former course is now silted up and covered by vegetation with garbage dumps located at points where old bridges allow ready access. Through examination of old maps and records and much careful fieldwork to gather accurate data on the new channel, geologist John Emerson has completed an assessment of the flood-related and geomorphic results of this early channelization project.

The new channel differs radically from the old one (Figure 8-4b). The present river is 29 kilometers (about 18 miles) long with a gradient of 3.1 meters per kilometer (about 16.4 feet per mile). The width of the constructed channel was originally 9

Figure 8-4. Stream changes resulting from the channelization of the Blackwater River. **a.** Dimensions before and **b.** dimensions after channelization. (Based upon data given by geologist John Emerson.) Note: Kilometers x .6214 = miles

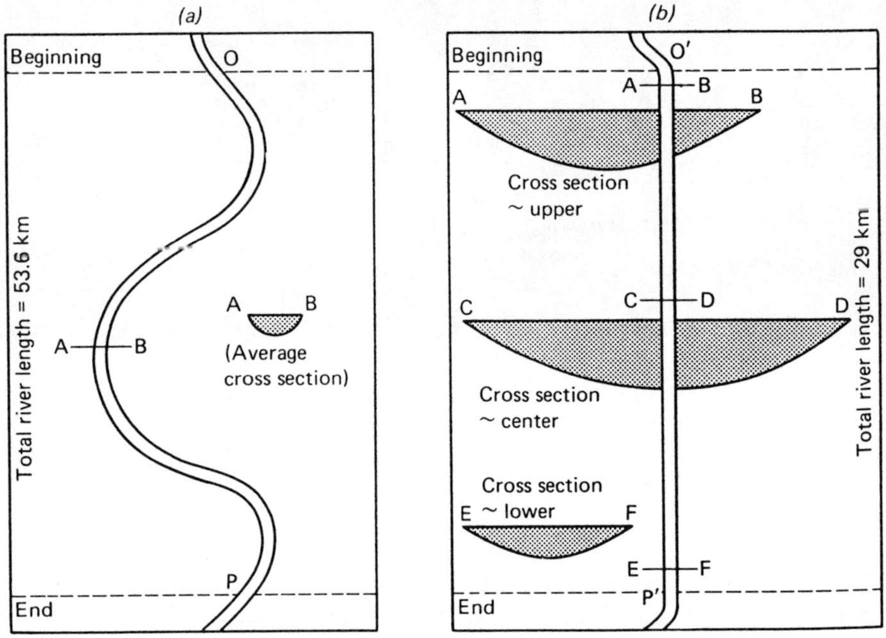

meters (about 30 feet) at the top, 1 meter (about 3.3 feet) at the bottom, with a depth of 3.8 meters (about 12.5 feet). The current dimensions are quite different. As the sketch shows, the channel has enlarged itself dramatically (Figure 8-4b).

The widening and deepening of the new channel since its construction is the result of serious erosion problems from bank cave-ins and the formation of erosion gullies. Emerson illustrates some of the problems associated with this change.

> A county bridge spanning the Blackwater had collapsed, owing to bank erosion, and was replaced with a chain bridge constructed by area farmers in 1930. The new bridge was 27 meters [about 90 feet] long. That bridge had to be replaced in 1942 and again in 1947. The final bridge had a span of 70 meters [about 230 feet], but it, too, has collapsed because of erosion [2: 325-326].

These changes are a result of the increased gradient of the channel. In turn, this new gradient modifies the amount of erosion and deposition until a new profile of equilibrium between the discharge and slope is established. Even in the sixty years since its construction no equilibrium was attained. When it is established, the new channel will resume a meandering course.

Consider, too, the extent of the channelization. The new channel does not extend to the confluence of the Blackwater and the Missouri. It stops at a boundary between two counties. The amount of water that can be contained within the channeled section is different from the amount that can be contained in the unchanneled part. Because of the abrupt change in the cross section at the boundary line, flooding has become a common occurrence at that location. As Emerson notes, longtime residents of the area have said that there was no flooding before the section of the river above them was channelized. Since that time, so frequently have floods occurred that two successive generations of fence posts have been buried by sediment in the floodplain.

As in the case of Crow Creek, the fish population of the channelized section has been modified. It is estimated that in the natural segments of the Blackwater the average fish content is approximately 630 kilograms of fish per hectare (about 565 pounds per acre). In the channelized section it is only about 125 kilograms per heactare (about 112 pounds per acre). We can see that while channelization of the Blackwater River allows more of the floodplain to be used in the channelized area, the erosion of farmland, the cost of bridge repair, and the increase in downstream flooding may well negate the other benefits.

The nature and effects of channelization in general are well summarized by Whipple, et al.:

> In the past, streams have often been modified to counter increasing flood peaks by straightening and lining natural channels so as to increase their hydraulic capacity. In many cases this has increased velocities sufficiently to increase stream erosion tendencies, has destroyed environmental values of natural streams, and has resulted in hastening floodwaters downstream, to the detriment of downstream flood-prone areas....The straightening and lining of channels should only be undertaken when the downstream effects of such actions have been identified and addressed [10:184].

LIVING ON A FLOODPLAIN

Geomorphology has acquired a highly specialized vocabulary. The early researchers used terms in their native languages to describe the new features they discovered, and these terms were adopted by later geomorphologists. Thus in the study of glaciation the German word Bergsrund is used to describe a type of crevice, and the French word arete identifies a knife-edge ridge. From Greek and Latin we have the word pedogenesis, meaning the formation of soil. There is one widely used English term, however, that perfectly describes a natural feature: the floodplain. As the name implies, it means a flattish area, a plain, that has been formed by floods (Figure 8-5).

It might seem obvious, given the very nature of the word floodplain, that anyone living on a floodplain should anticipate floods (Figure 8-6). Indeed, without floods there would be no floodplain. However, today with the growth of the population and the resulting building development, floodplains have been extensively developed in many areas of the world. We can read accounts in newspapers regularly of disasters in floodplain areas and loss of homes, factories, and even lives.

Figure 8-5. Floodplain development: (a) A mature river with a floodplain almost wide enough for its meanders. (b) A river in old age with a broad floodplain and freely developed meanders. (From A.N. Strahler, Physical Geography, 4th ed. (1975), p. 420. Courtesy of John Wiley and Sons, Inc.)

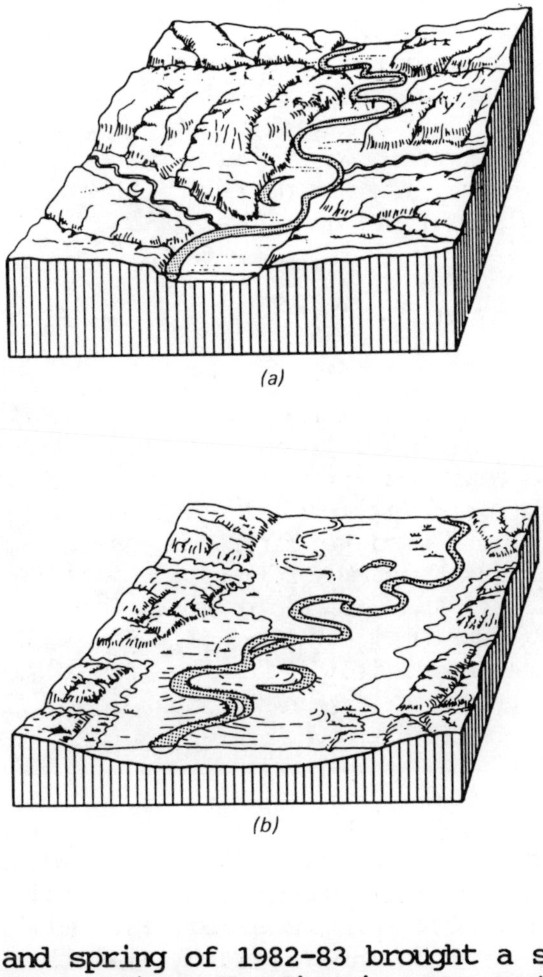

(a) (b)

The winter and spring of 1982-83 brought a series of severe storms that led to serious flooding in many parts of the United States. In April Louisiana, Mississippi and Alabama were especially hard hit. Rainfall of up to 300 millimeters (twelve inches) caused massive river flooding, leaving seventeen people

Figure 8-6. Construction of homes on sites that "belong" to the river leads to scenes such as this. (Photo courtesy of U.S.D.A., Soil Conservation Service).

dead and 52,000 homeless. Loss estimates exceeded $600 million. In May storms again brought disaster to lowlands, especially in Mississippi, where Jackson received 180 millimeters (seven inches) of rain in a four-day period, and the flooding Pearl River put a fifth of the city under water. Elsewhere in the state virtually every stream was flooded, and water covered half a million hectares (about 1.2 million acres) of land. Nearly 6,000 residents were evacuated [6:10].

Such catastrophes are not so unusual, and we should question why the damage on floodplains reaches such vast proportions. In part the answer may be related to the flood-prevention precautions instituted by numerous government agencies. Figure 8-7 shows how small floods can be contained by dikes, perhaps giving the residents of such an area a false sense of security. In Figure 8-7a we see that under natural conditions the floodplain is inundated at intervals of about five years. Levees are constructed (Figure 8-7b), and the five-year flood no longer extends onto the floodplain. With fewer floods land development occurs. Eventually, however, a flood of great magnitude will rise higher than the levees and cause great damage and financial loss (Figure 8-7c). Ironically, if the protective structures had not been there, the cost of the flood would not have been so large. With the control measures the cost was much greater.

PERCEPTION OF FLOOD HAZARD

Now consider the people living in the flooded area. They may well have lost all their possessions. Will they reoccupy their former homes? If they do, why do they do it? A number of researchers have conducted studies on flood perception to try to answer these important questions. They obtained some surprising results.

In a report published in 1961 Wolf Roder examined the attitude of residents living on the floodplain of the Kansas River at Topeka [9:62-83]. Since 1903 the river had overflowed its banks sixteen times; so it represented an ideal location for studying attitudes toward living on a floodplain. Furthermore, because of these floods, extensive flood-prevention measures had been installed so that the study also allowed assessment of the false sense of security that floodplain residents had. Roder attempted to determine whether people were optimistic or pessimistic in their attitude toward flooding. He wished to test the hypothesis that the government construction

Figure 8-7. Schematic representation illustrating how the flood control measures may encourage settlement on dangerous land.

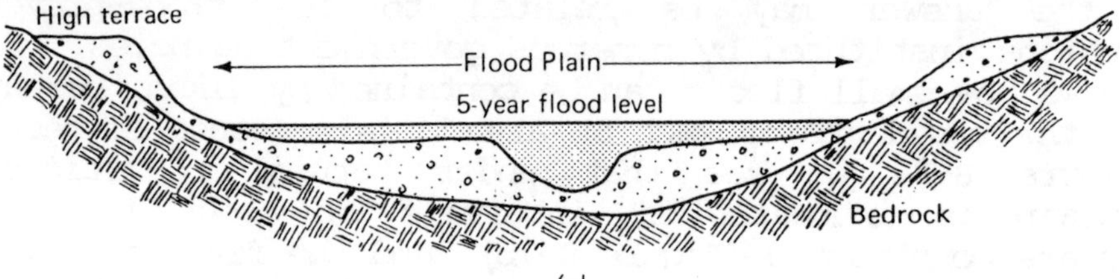

(a)

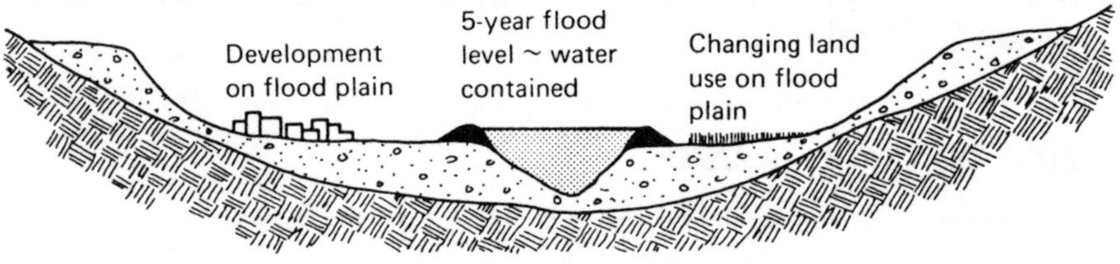

(b)

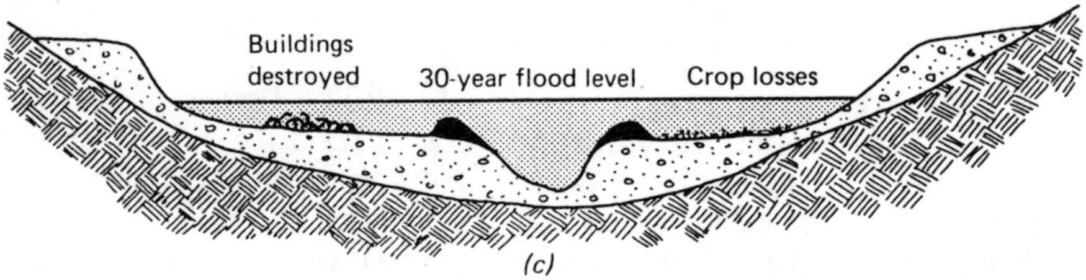

(c)

of levees and dams influenced the decision to live on a floodplain. In order for the preventive measures to be a factor, the residents had to be aware of the flood-control activity. Roder found that of the persons he interviewed, some 30 percent had no knowledge whatsoever of protective structures. Yet 63 percent of these people were optimistic in their attitude toward future flooding. This response is only a little less optimistic than the attitude of people who knew about the structures (19 percent of those interviewed), of whom 70 percent were optimistic. It seems that while protective measures may increase optimism, the individual is more likely to be influenced by experience than by specific information about actions taken. It is also significant that knowledge of the construction of flood-control devices is largely confined to people in the upper socioeconomic levels.

Another phase of the investigation concerned the attitudes held by persons who had actually suffered losses through a serious flood. It might be anticipated that having experienced a flood disaster, people would hesitate about relocating on a floodplain. Roder found, however, that 20 percent of the people interviewed had actually been through a flood before, some in the same locale, others on other floodplains. The fact that they relocated their homes on a floodplain could not be simply explained by low-cost housing. Although rentals in the floodplain were lower than most others in Topeka, there was housing available at competitive prices. Therefore, their decision was not simply a case of having nowhere else to go.

What, then, are the incentives that cause people to live--or to relocate after a flood--on a floodplain? The question must be qualified at once because some people do not remain. In Topeka many persons who rented houses had little reason to stay and so moved. Similarly, wealthier people, who could absorb the financial loss, also left. Roder found that the major reason for staying was that many people were unwilling to leave their friends, relatives, and familiar surroundings. "I wouldn't live anywhere else." "This town is my home. All my friends are here." It seems that when they evaluate the desirability or undesirability of a place to live, they give the possibility of flooding a low priority.

Following Roder's research, Ian Burton conducted a similar study in the floodplain of the Little Calumet River in northern Indiana [1:84-92]. Unlike the Topeka case, no extensive flood protection measures had been constructed nor were any projected. Burton's data are given in Table 8-2. In that table, part (a) shows the extent of housing development on the floodplain. The data include statistics for the towns of Hammond and Munster, about one and a half kilometers (one mile) from the river. A phenomenal development is apparent. Part (b) lists two of the three times when flooding occurred. Another flood took place in 1954 when the peak discharge was 126 cubic meters (2,600 cubic feet) per second. The data clearly indicate that the amount of damage is closely related to the development on the floodplain. The 1957 data would have been appreciably larger had not some flood-protection work been completed subsequent to the 1954 flood, when damage was estimated at $823,400. Part (c) summarizes the attitudes toward floods. About half of the respondents to the study on the Little Calumet had had some experience with floods, and most of the others lived in homes built subsequent to the 1957 flood. Those persons who had not experienced floods were appreciably more optimistic about the possibility of future floods. Nonetheless, a surprisingly large number of flood victims were also optimistic about the future. Burton notes that such a viewpoint depends on the perceived severity of the flooding. To some, "water lapping at the doorstep and several feet deep in the basement is 'not a real flood'." When a severe flood took place, their attitude bacame more pessimistic.

The two surveys both point to the complexity of attitudes toward flooding. As with any of the physical hazards that people face, there is no simple basic set of rules or guidelines. Since people have their own goals and desires, a group response can never exactly represent individual attitudes. Nonetheless, since the completion of these two early studies, many other surveys of the human response to flooding have substantiated these findings. There are particularly good examples of these attitudes in Natural Hazards [11], which provides case studies on attitudes toward floods in areas

Table 8-2. Attitude Toward Flooding

(a) Development of Housing on Little Calumet Floodplain

Year	No. of Houses
1940	140
1945	334
1950	652
1955	2030
1960	3928

(b) Examples of Floods and Resulting Damage. Note: cubic feet 20.6415= cubic meters.

Year	Peak Discharge (in cubic feet per second)	No. of Houses Damaged	Estimated Cost of Damage ($)
1947	2490	7	20,200
1957	2020	278	161,600

(c) Attitude Toward Future Flooding

	Response from Persons Not Having Experienced the Earlier Floods	Response from Persons Not Having Had Experienced Earlier Flooding to a Minor Degree
Percent pessimistic	11	11
Percent neutral	25	46
Percent optimistic	64	43

Source: Burton [1].

ranging from Shrewsbury, England to Opotiki, New Zealand. In some of the studies earlier findings were corroborated; in others new light was thrown on the topic.

One particularly interesting observation came from a study of floods on the Rock River in the vicinity of Rock Island and Moline, Illinois. Investigators completed two attitude and perception surveys, one in 1970 and a second a year later. Between these two times the study area experienced serious and costly flooding. Much to the surprise of the researchers, attitudes observed before and after the flood showed few differences. For example, the 1970 survey found that a clear majority (65 percent) of the sample emphasized the advantages rather than the disadvantages of living in the area. Despite the recent floods, the 1971 survey found the same response, with 69 percent stressing the advantages of their location. Given such results, it is evident that interpretation of human attitudes needs to be based on much more than actual physical events. The study of flood attitudes, like the study of the perception of any natural hazard, remains an area of intensive, environmental research.

CONCLUSION

In summary we must conclude that for each of the three problems we have discussed there is no clear answer. The methods used for identifying flood recurrence intervals are far from perfect. The problems that result from channelization may well outweigh any flood-prevention benefits. And finally, the general disregard for the hazards of living on floodplains is evident from the survey results. Although the picture is rather bleak, there have been positive developments: watershed planning to modify the amount of water that goes into streams, and zoning ordinances to limit the use of floodplains. Nonetheless, there is still much work to do if we are to decrease the number of flood-related disasters.

SUGGESTIONS FOR FURTHER WORK

1. Drawing upon local newspaper accounts over the last five years, describe and explain any floods that took place in your area. If estimates of damage are given, find the areas most affected and ascertain if new developments have taken place since the flood.
2. Using reports published in various sources, list at least three flood disasters that have recently occurred. For each of these explain the cause of the flooding and why loss of life resulted.

SELECTED SOURCES FOR FURTHER READING

Davis, Kenneth S. River on the Rampage, Garden City, N.Y.: Doubleday and Co., 1953.

Hockin David L. "Channelization: Some Engineering Aspects", The Geographical Journal, 151 (March, 1985),pp. 54-57.

McComb, David G. Big Thompson: Profile of a Natural Disaster, Boulder: Pruett Publishing Co., 1980.

Smith, G.H. (ed.). Conservation of Natural Resources, 4th ed., New York: Wiley, 1971. See Chapter 13.

Ward, Roy. Floods: A Geographical Persepctive, London: Macmillan, 1978.

White, Gilbert F. Flood Hazard in the United States: A Research Assessment, Boulder: Institute of Behavioral Science, University of Colorado, 1975.

REFERENCES

1. Burton, Ian. "Invasion and Escape on the Little Calumet," in G.F. White, ed. Papers on Flood Problems, Chicago: University of Chicago, Department of Geography, Research Paper No. 70, 1961.
2. Emerson, J.W. "Channelization: A Case Study," Science 173, 1971.
3. Gillette, R. "Crow Creek: Case History of an Ecological Disaster", Science, 176, 1972.
4. Helley, L.J., and V.C. Lamarche. December 1964, A 400-Year Flood in Northern California, Washington: U.S. Geological Survey Professional Paper 600D, 1968.
5. Hoyt, W.G. and W.B. Langbein. Floods, Princeton: Princeton University Press, 1955.
6. LeComte, Douglas. "The Weather of 1983", Weatherwise 37: No. 1, February, 1984.
7. Leopold, L.R. Water: A Primer, San Francisco: W.H. Freeman, 1974.
8. New England River Basins Commission. The River's Reach: A Unified Program for Flood Plain Management in the Connecticut River Basin, Hanover, N.H.: New England River Basins Commission, 1976.
9. Roder, Wolf. "Attitudes and Knowledge on the Topeka Flood Plain," in G.F. White, ed. Papers on Flood Problems, Chicago: University of Chicago Department of Geography, Research Paper No. 70, 1961.
10. Whipple, William, Neil S. Grigg, et al. Stormwater Management in Urbanizing Areas, Englewood Cliffs: Prentice-Hall, 1983.
11. White, G.F. Natural Hazards: Local, National, Global, London: Oxford University Press, 1974.

Chapter 9

The Aswan High Dam

The Context

A number of social scientists have suggested that the early development of societies depended upon the control of water for irrigation purposes. The partial control of the Hwang Ho River in China and the Eurphrates River in Mesopotamia (modern Iraq) are frequently cited as examples, Although the close relationship between the development of water control is arguable, the evidence certainly shows that human beings have long depended upon water control for agriculture.

Since those early times, people have continued their efforts to control natural water supplies. Today, with the growth of population, new energy needs, and intensified agriculture, water control systems using dams require vast engineering projects. The size of modern dams--the Nurek Dam in the Soviet Union is 309.9 meters (1,017 feet) high, and the Tarbela Dam in Pakistan is composed of 141 million cubic meters (186 million cubic yards) of material--indicates the technological advances in water-control constructions.

Although engineering is the main technological concern in dam construction, environmental and social sciences play a role in assessing the dam's eventual impact on the environment. Unfortunately, a thorough understanding of side effects of dam construction is not always immediately available. Nor are the side effects always anticipated. In this chapter we shall consider some of these effects in relation to the Aswan High Dam in Egypt, and its reservoir, Lake Nasser.

Most of us are aware that there is a great deal of political and economic maneuvering associated with resource development, and the Aswan High Dam was no exception. In the early to mid-1950's, World Bank, USAID (United States Agency for International Development), and Egyptian scientists undertook a series of engineering and site studies on the feasibility of building a major storage facility at Aswan. Recognition of communist China in 1956 by President Nasser led to a withdrawal of promised financing by the World Bank, Britain, and the United States. Nasser retaliated by nationalizing the Suez Canal and a political crisis resulted. Ultimately, the Egyptian government sought funding and technical assistance from the Soviet Union, and the Aswan project was begun in 1960 and completed in 1967. The alliance of these two countries was no small achievement for Egypt's President Nasser (Figure 9-1).

The High Dam was intended to accomplish a number of goals. Water stored in its 380-kilometer (300-mile)-long reservoir would provide additional water for irrigation and a new source of protein from fisheries along its banks. As a flood-control

Figure 9-1. The Aswan High Dam. (Courtesy of United Nations).

measure, the dam would prevent severe floods in the lower reaches of the Nile and, by judicious release of stored water, would avoid the drought periods that sometimes occurred. Regular release of water could be used for supplemental irrigation and other purposes. Further, the dam would generate 10 billion kilowatts of electricity a year to supply power for the developing industries of Egypt. Ecologically, the dam created some severe problems that were both expected and unexpected.

Figure 9-2 shows the location of the High Dam. It is about eight kilometers (five miles) south of the Aswan Dam, which was constructed in 1902 at the first Nile cataract in Egyptian territory. The High Dam dwarfed it. The High Dam is about 4.2 kilometers (2.6 miles) long, about 105 meters (346 feet) high, and approximately 1,200 meters (3,900 feet) thick at its base.

Figure 9-2. Location of the Aswan High Dam. Inset graphs show river discharges before the dam was constructed, in hundred million cubic meters per day. (Note: One cubic meter equals 264.2 U.S. gallons). (Redrawn from W.A. Hance, The Geography of Modern Africa, 1964, p. 123. Reproduced by permission of Columbia University Press.)

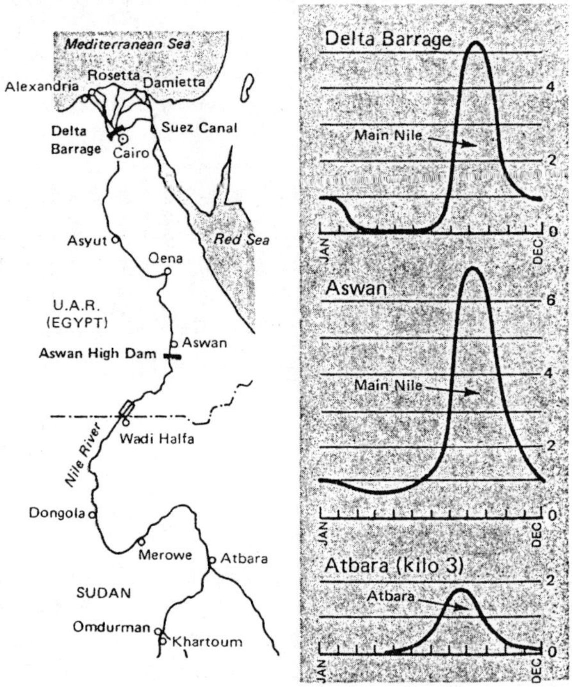

Its reservoir, Lake Nasser, stores 162 billion cubic meters (about 132 million acre feet) of water and occupies some 5,560 square kilometers (about 2,000 square miles).

The impact of the High Dam far exceeds the goals for which it was constructed. Its effects are seen in the Nile itself, in the people who live along the river, in their economic practices, in the quality of the land they farm, and in changes in the long-term ecological relationships between the river, the land, and the people.

THE NILE AND THE RESERVOIR

The River Nile is the essence of Egypt. Its source is in the mountains of equatorial Africa, and its flow is augmented by rivers descending from the Ethiopian Highlands. The river is large enough to maintain its flow through the desert that comprises Egypt. In the final 900 miles before it reaches the sea, no tributaries flow into the river. Therefore, the level reflects the amount of rainfall in the source area. The Nile rises rapidly in May and reaches flood stage in September. Thereafter, its level subsides to reach its lowest stage in April. Patterns of the average discharge curves, prior to the construction of the High Dam, are shown in Figure 9-2; however, discharge has been highly variable from year to year.

Since the time of ancient Egypt, farmers living along the Nile have irrigated their lands, using the floodwaters. Carrying silt, the water flooded the fields, which were surrounded by dikes, and the silt settled from the water to form new soil. When floods were high, there were good crops. When there was less water than usual, famines sometimes occurred.

One purpose of the High Dam, like most dams, is to regulate the flow of water to the lower reaches of the river. In modern times before the building of the dam, annual flow varied from a low of 45 billion cubic meters (165 billion cubic yards) to a high of 150 billion cubic meters (196 billion cubic yards). Since closure, the highest flow into the reservoir has been 125 billion cubic meters (165 billion cubic yards), and the lowest just over 40 billion (52 billion cubic yards) in 1983, with releases from the dam averaging about 55 billion cubic meters (70 billion cubic yards) yearly. A dry cycle in the 1980's has resulted in low flows and a lowered reservoir level so that if another dry year or two occurs, it may be necessary to curtail irrigation and power generation. The object is to regulate the great fluctuations of flow by storing floodwater and releasing it during times when the river is normally low. In order to maintain a required level, it was estimated that Lake Nasser would need to contain 163 billion cubic meters (about 132 million acre feet) of water. While the lake has such a capacity, there was debate on how rapidly the reservoir would fill, if ever. Except for one great flood in 1975, drought occurred in the upper reaches of the Nile, following closure. By 1977, the reservoir had filled to a level of 177 meters (about 580 feet). Engineers originally estimated that in this hot, dry climate some 10 billion cubic meters (13 billion cubic yards) would be lost through evaporation each year. Current data support the initial estimates of about 11 percent reduction of lake volume annually by evaporation. Loss of such a vast amount of water means that less is available below the dam.

The seemingly simple process of holding water to regulate river flow is not as straight forward as it appears, for the modifications have had many side effects. First, we must consider the geomorphological problem. The Nile, like all rivers, has evolved a graded profile through time (Figure 9-3a). This profile means that the slope of the river and its channel characteristics are adjusted so that the average volume of water passing through it has the right velocity to transport the load supplied from its drainage basin. A graded river is not cutting deeper into its valley. It has attained a state of equilibrium. By modifying any of its characteristics, such as the volume of water to the load it carries, a new equilibrium needs to be established (see Figure 9-3b). Dams create such modifications.

The water passing below the dam does not carry sediment because it has settled in the reservoir. The clear water, flowing under a modified regime, now has the ability to scour or erode the channel below the dam. The study of this effect in United States rivers shows significant quantities of materials are eroded (Table 9-1). In the Nile Valley, a channel below the dam which may have been stable for hundreds of years, was expected to show similar effects. Since closure, channel erosion has ranged from 250-700 millimeters (about 10-28 inches) in several stretches of the river and has been significant enough to warrant the construction of concrete weirs below downstream dams, protective structures for some navigation channels, and revision of standards for channel construction [7:10].

Figure 9-3. The graded profile of a stream. a) The typical form. b) An example of how interruption of equilibrium results in the regrading of the bed. This example shows how a waterfall, created by an earthquake on Cabin Creek in Montana, was regraded in a period of 3 years.

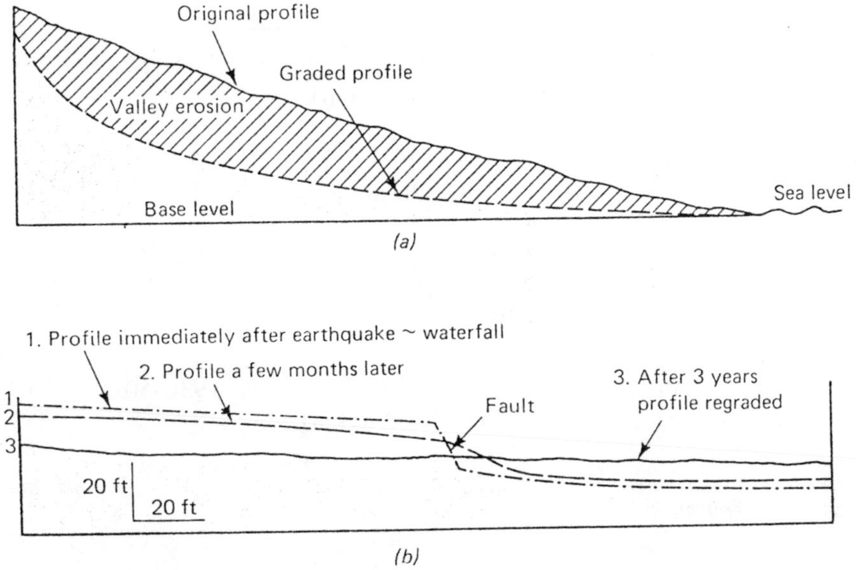

The sediment deposited in Lake Nasser will, eventually, destroy it. Again, lakes in the United States that are formed by dams illustrate the problem. [2] Some reservoirs constructed in the western states in the early 1900's have already been completely filled by sedimentation.

The depositing of sediment in the reservoir has also had geomorphic effects in the Nile delta area. Deltas are built when the load of a river is deposited at a point where the river leaves its channel and enters a larger body of water--in the case of the Nile the Mediterranean Sea. The formation of the delta depends upon many factors, but one is the amount of

Table 9-1. Erosion Below Selected Dams

Dam	River	Drainage Area (sq miles)	Period of Record	Average Lowering of Bed at All Sections (ft)
Guarnsey	N. Platte (Wyoming)	16,200 (41,958 sq km)	1927-57	2.00 (0.60 m)
Canton	N. Canadian (Oklahoma)	12,483 (32,330 sq km)	1947-59	1.82 (0.55 m)
Denison	Red (Oklahoma-Texas)	38,291 (99,432 sq km)	1942-48	1.63 (0.48 m)
Fort Peck	Missouri (Montana)	57,725 (149,507 sq km)	1936-50	1.01 (0.30 m)

Source: from Fluvial Processes in Geomorphology *by L. G. Leopold, M. G. Wolman, and J. P. Miller, W. H. Freeman and Company. Copyright ©1964.*

sediment available. Since the construction of the High Dam, the rate of erosion along the Mediterranean coast around the delta has increased with the loss of silt deposits from the Nile.

The Nile carried approximately 54-162 million metric tons (60-180 million U.S. tons) of silt, depending on variations in flow, and it was recognized that most of this sediment would be deposited in Lake Nasser when stream velocity was reduced. As reservoir filling progressed, sediment deposition would progress up the reservoir, and delta building would begin in the reservoir. Silt accumulation has averaged 100 million metric tons (about 110 million U.S. tons) annually, and at present rates will take approximately 500 years to fill the dead storage capacity of the reservoir. A major unknown variable is the impact that potential land use changes in the upstream watershed might have on the rates of sedimentation.

Coastal erosion appears to be severe in only a few delta sectors; however, rates of 25 meters (approximately 82 feet) per year are occurring at the Rosetta headland.

FARMING, FISHING, AND SEDIMENT

Some of the sediment that settles to the bottom of Lake Nasser used to be deposited on the banks and in the diked fields of the Egyptian farmer. The sediment added nutrients to the soil and was often noted as being a major source of soil fertility. The silt also improved the tilth of the naturally sandy soils. Without this annual addition of sediments, the fertility and tilth of the soil has declined, and to maintain food production, chemical fertilizers have been added.

Today, calcium nitrate produced at Aswan, using power from the dam, provides a source of fertilizer for agricultural production. Unfortunately, many agronomists feel that land productivity is not being maintained despite greater technical inputs into the agricultural sector. Deterioration of soil fertility has resulted not only from the absence of river silt, but also from more intensive use of the land through multiple cropping, the application of chemicals, and poor cropping techniques.

Salinization of the soil is a problem faced by irrigation projects in all arid regions. As described in the Pakistan example (see Chapter 10), salinization leads to a reduction in crop yield and eventually to the loss of arable land. Salinization has also been a problem for many years in the waterlogged delta of the Nile. Today, it is spreading to other irrigated areas.

In order to combat the threat of salt accumulation in the soils, a program of tile drainage, improved drainage ditches, and pumps was initiated to move water through the irrigated soils and out of the saturated areas. This program has proceeded slowly. Unfortunately, in a number of cases the design of materials was unsuitable, and large-scale projects have not functioned as expected. The salinization-drainage problem is very complicated, and it is at present difficult to assess the impact of drainage on the movement of nutrients and salts. What is clear is that perennial, rather that flood, irrigation coupled with multiple cropping and reduced closing of canals for cleaning and repair has extended the periods of soil saturation in the root zone and limited the period of aeration appreciably. A relatively recent FAO (Food and Agriculture Organization) study estimated that one-third of the nation's agricultural land has problems with salinity, while ninety percent of the land is affected by waterlogging. Thus, the additional cost of dealing with the salinization created by the dam is forced upon the economy.

Rapid population growth was one of the major motivating factors in the government's undertaking the construction of the dam. At the national level was the urgent problem of feeding Egypt's population, which has grown from 26 million in 1960 to 51 million today. Expansion of irrigated acreage, the use of multiple cropping (two crops per year are relatively common and three crops are possible in many areas), and the application of modern technology were expected to solve the problem. Despite multiple cropping and the application of some of the highest amounts of fertilizer in the world, the problem of meeting the nation's food needs has not been overcome. Food imports, which have risen from over $100 million in the early 1970's to more than $3 billion in the 1980's, are in part responsible for making Egypt one of the largest debtor nations in the region.

The envisioned increase of irrigated land also has not taken place. A number of factors are responsible for the amount of irrigated land remaining virtually unchanged since the inception of the dam. Figures differ from source to source, but Susan Walton estimates that:

> In addition to the acreage converted to year-round cultivation, about 950,000 acres [350,000 hectares] of desert land have been reclaimed for agriculture. But since this gain is partially cancelled out by the loss of 600,000 acres [about 243,000 hectares] to urban expansion -- and by additional acreage lost to shoreline erosion along the Mediterranean -- the net gain of agricultural land is slight [5:32].

In spite of the thousands of hectares of new irrigated land being developed at tremendous costs, Egypt has continued to lose to urban and industrial sprawl an amount of land equal to that gained by the dam construction. The situation is, in fact, much worse than the figures indicate, as the quality of land involved is not taken into account. Much of the land lost to urbanization is high quality, while much of the reclaimed land is of marginal quality. Thus, the reclaimed land requires special soil treatment at relatively high costs in order to obtain moderate yields. Additionally, it has taken longer to develop these poorer soils than expected and has required more water than anticipated, all at a considerable cost. Some of the land has succumbed to salinization and waterlogging, and other large parcels have returned to nature.

Regulated discharge led to other changes on the Nile banks and adjacent lands. Bank erosion has increased downstream from the dam. More importantly, deprived of the annual increments of silt on the banks of the river, brick making operations expanded to obtain needed supplies of clay. To obtain raw materials for their operations, brick makers purchased agricultural land, which was excavated to the depth of the water table. Approximately 1,000 square kilometers (approximately 386 square miles) of agricultural land had been lost by 1984 when legislation was enacted to prohibit this activity [7:34].

Some fishermen, too, are feeling the impact of the dam. Downstream from the dam both the number of fish and the species of fish caught have declined. Because of the sediments deposited in Lake Nasser, water reaching the Mediterranean lacks the organic nutrients which travel in the sediment. Marine organisms, particularly sardines and shrimp, are deprived of an adequate food supply. As a result, Egypt's sardine industry, which used to produce some 7,300 metric tons (about 8,000 U.S. tons) of fish annually, has virtually disappeared; and catches of shrimp have decreased by about one-third. Demersal fish catches have decreased by approximately two-thirds. While overfishing is involved, there is good reason to believe that changes in water quality and nutrients are significant factors in declining fishery production.

In 1965, projected fish production from Lake Nasser was in the range of 14,500-18,000 metric tons (16,000-20,000 U.S. tons) although some government experts claimed that the lake would eventually produce 54,000 metric tons (60,000 U.S. tons) of fish annually. Early production was about 4,500 metric tons (5,000 U.S. tons); but by 1978 the catch had exceeded 20,000 metric tons (22,000 U.S. tons), amid growing concern about overfishing. There are also problems related to the management of the operation, the welfare of the fishermen, and the lack of transport to the major market in Cairo.

Hopefully, the fishing industry in Lake Nasser will not follow the pattern of Zambia's Lake Kariba. There, the 18,000 metric ton (20,000 U.S. ton) projected annual yield seemed at first easily attained, but a few years later the catch was only 3,600 metric tons (4,000 U.S. tons), and by 1978 the fishing industry was practically nonexistent [4]. Production has rebounded to about 10,000 metric tons (11,000 U.S. tons), based largely on the introduction of new species; however, this tonnage is still far short of the original projections [3]. Development of fishing villages on Kariba have led to other environmental problems. Tree cutting to satisfy local wood-fuel needs and to export commercially to urban areas has caused deforestation. This, in turn, has led to conflict with the National Parks Policy, to maintain wildlife reserves. The ecology of these new lakes is highly complex, and we have much to learn about successive processes that occur.

DISEASE

As might be anticipated, the reduction of the floods that formerly flushed the areas bordering the Nile has created health hazards. Rodent populations have increased, and sewage systems, once cleaned by the rising and falling of the river, have become badly clogged. But the health problems associated with these results are dwarfed by another disease: the parasitic disease known as bilharzia or schistosomiasis. As John Weir has written:

> Schistosomiasis is the greatest unconquered parasitic disease now afflicting men and animals. It is a disease of rural people, and is a man-made disease resulting from the lack of sanitary facilities. Tragically, it is growing in seriousness in part because the canals and ditches needed for farming serve as breeding places for snails that are vectors of the parasite [6].

The disease is caused by parasitic flatworms, known as schistosomes. Humans are infected by three common species: S. Haematobium, S. Mansomi, and S. Japoni. Figure 9-4 shows the schistosome life cycle. The cycle begins when a cloud of larvae (cercariae) leaves a host snail that is in the water. The cercariae then penetrate the skin of animals and humans and grow to adult flukes, or worms, in veins or venules around the small intestine, where they lay their eggs. Some of these eggs will migrate to various organs such as the brain, spinal cord, lungs, bladder, appendix, rectum, spleen, and liver. If a mature egg falls into the water, it will rapidly produce a larva or miracidium. The larva then seeks a host snail to start a new cycle.

In the human the eggs of the worm pass freely through the blood stream, often in vast quantities. As many as 100,000 eggs have been found in the lungs of persons severely afflicted. Eggs can continue to be produced by a worm for as long as 20 years, and the disease becomes progressively worse. Initially,

Figure 9-4. The schistosome life cycle in human beings and snails. (Adapted from J.M. Weir, "The Unconquered Plague," Rockefeller Foundation Quarterly No. 2, 1969. Reproduced by permission.)

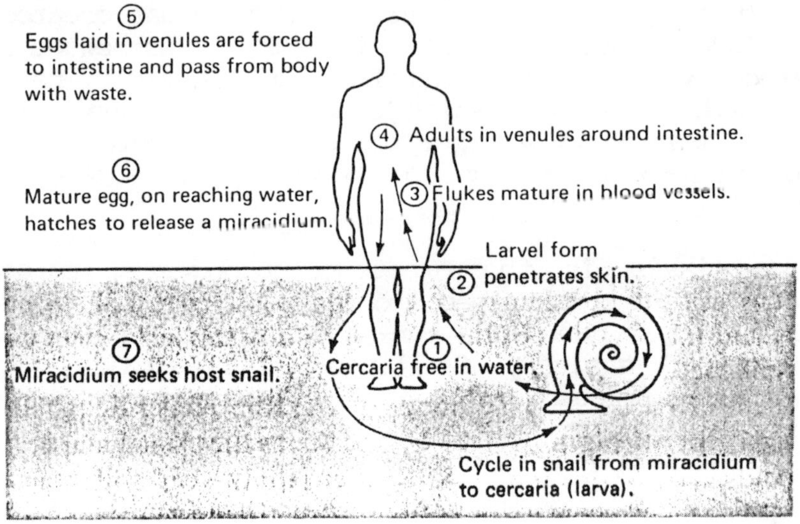

the effects may not be observed, for the disease only reaches the disabling phase at a time when normal degeneration through aging occurs. The chronic sufferer faces a life of increasing exhaustion, often being unable to work for more than a few hours a day, and increasing pain.

Schistosomiasis is not a new disease in Egypt. Eggs have been found in mummies dated at 1,000 B.C. Even before construction of the Aswan High Dam, it was estimated that one out of every four persons suffered from schistosomiasis. Projects such as the dam create ideal habitats for both snails and the schistosome parasite. The snail carrier cannot exist in fast-flowing water but is at home in the 300-mile placid water shoreline of the lake and canals. As a result of altered water quality and perennial irrigation, hydrophytic vegetation has expanded and diversified; and thus the snails' habitat has been extended. Of course, chemical treatment and mechanical clearing of the canals are possible, but to control aquatic vegetation represents an additional cost. Although no official data are available about the change in the frequency of schistosomiasis since the dam was constructed, scattered studies on the disease suggest that urinary schistosomiasis has decreased in village populations. This decrease is probably related to better village water supplies, better health care and education, and prophylactic treatment. Intestinal schistosomiasis, however, has increased; and although the causes are not clear, they may be related to the dam. What is known is that the prevalence of gastrointestinal infections continues to be high.

The eradication of schistosomiasis is a difficult task. At the base of the problem is sanitation. The schistosome life cycle can only be maintained when the same water is used for everything from washing to sanitary-waste disposal, a common procedure in the Nile Valley. To alter a life style requires far-reaching changes in social habits and customs, a change not easily implemented. Given the fact that sanitation improvement takes time, it is not immediately possible to eliminate schistosomiasis; therefore, other approaches have been tried. Drugs are available for treatment of individuals with the disease; but because the treatment requires hospitalization and can have unpleasant side effects, it is not really effective for mass treatment. Killing the host snails has also been attempted, but application of the appropriate chemicals is costly, affects the fish population, and clearly cannot be used in water that people drink.

The search thus continues for a solution. Other chemicals for the elimination of the snail are being tested, and researchers are seeking a vaccine for immunization against the disease. Despite such efforts, there is little doubt that the numbers affected by schistosomiasis will continue to remain high before a solution is found.

CONCLUSION

We have discussed some of the ways in which the Aswan High Dam has had an impact on the people of the Nile Valley. Although some of these changes were anticipated, others were unexpected. As time passes, still other environmental surprises may be in the offing.

To assess the full meaning of the Aswan High Dam in Egypt, all of the effects described in the preceding pages need to be weighted against the benefits that the dam provides. Because we have emphasized the environmental side effects, we have undoubtedly given the impression that the dam is one huge disaster, but such a conclusion may not be warranted. For example, 1971 and 1972 could well have been "lean years" for agricultural production had not the water in Lake Nasser been used to alleviate the exceptionally dry conditions. Similarly, while benefits have not been spread evenly, some of the farmers of the Nile Valley have reaped benefits from the construction of the dam. Richard Critchfield, who has studied the peasant life and culture in Egypt, cites the opinions of a number of fellahin, or Egyptian peasants. One of these, who had profited from the land reform acts of the 1950s and now owns 35 acres, said:

> In the past I had 100 acres. I grew lentils, wheat, anything. I could get 1,000 pounds (one pound being worth $1.75) from those 100 acres and I paid 200 pounds for seeds. So I got 800 pounds profit. And perhaps from those 100 acres, 20 to 30 might be lost to insects. Now if I cultivate sugar cane I invest 7000 pounds, but I get back perhaps 15,000 on 100 acres; more than half is profit and I get much money. Before the High Dam, I only sowed one crop a year; now I get three [1].

Gilbert White believes that we now know enough about the environmental effects of the Aswan High Dam to identify and begin to assess the project's impact. In relation to the major stated goals, the project largely achieved its purpose; however, irrigation acreage expansion has not occurred. Most of the environmental side effects were anticipated in advance, with greater or lesser accuracy. One major exception was the change in water quality and the availability of nutrients in cultivated fields which may be of critical importance in the future.

Public perception often holds large dams to be of great social and economic benefit because of flood protection, power production, and provision of water for irrigation. Some scientists argue that large water projects and dams not only have often failed to achieve their objectives, but also have given us a legacy of environmental damage and cultural destruction that have resulted in a staggering array of problems that remain unsolved. The Aswan High Dam exemplifies some of these issues.

SUGGESTIONS FOR FURTHER WORK

1. As mentione din this chapter, the dam that has formed Lake Kariba has created ecological problems. What problems have been created and how do they differ from those of the Aswan High Dam?

2. Consider some of the benefits and problems of the following dams and their reservoirs: Elephant Butte Dam, th.

SELECTED SOURCES FOR FURTHER READING

Clements, T. "St. Francis Dam Failure of 1928," in R. Tank, Focus on Environmental Geology. New York: Oxford University Press, (1973), pp. 460-63.

Goldsmith, Edward and Nicholas Hildyard. The Social and Environmental Effects of Large Dams Vol. 1. Cornwall: Wadebridge Ecological Centre, 1984.

Hilton, T.E. "The Akosombo Dam and the Volta River Project," Geography, (1966) 51: 251-54.

Man and His Geologic Environment, Reading. Massachusetts: Addison-Wesley, (1974), pp. 415-16.

Morgan, A.E. Dams and Other Disasters. Boston: Porter Sargent Publishers, 1971.

Reeve, W.H. "The Kariba Dam," Geography Journal (1960) 126: 140-46.

Robin, Neville and William M. Warren. Dams in Africa: An Intre-Disciplinary Study of Man-Made Lakes in Africa. Belfast: W&G Baird Ltd., 1968.

REFERENCES

1. Critchfield, R. Los Angeles Times, September 3, 1975.
2. Leopold, L.B., M.G. Wolman, and J.P. Miller. Fluvial Processes in Geomorphology. San Francisco: W.H. Freeman, 1964.
3. Magadza, C.H.D. "Conflicts of Resource Use on the Lake Kariba Environs," October-December, 1986.
4. Scudder, T. "Kariba Dam: Ecological Hazards of Makin a Lake," Natural History, Vol. 78, No. 2 (1969), pp. 68 ff.
5. Walton, Susan. "Egypt After the Aswan Dam," Environment. Vol. 23, (1981), No. 4.
6. Weir, J.M., "The Unconquered Plague," Rockefeller Foundation Quarterly, No. 2 (1969), pp. 4 ff.
7. White, Gilbert F. "The Environmental Effects of the High Dam at Aswan," Environment, Vol. 30, No. 7, (1988), pp. 5-40.

Chapter 10

Irrigation:
The Problem of Salinization

The Context

The two photographs on the facing page illustrate two quite different ways of irrigating land. One shows a method that has been used for thousands of years. The other shows a system using modern technology. Despite the different levels of technology in the two systems, they both face similar problems: when to add water to the land and how much to add. In the primitive system past experience probably provided the answer. In the modern system the relationship between the amount of water

a. "Ancient"

(a)

Figure 10-1. Irigation methods, a. ancient (Photo by Barry Michie) and b. modern (Photo by Stephen Stover).

b. "Modern"

(b)

used by the crops and the prevailing climatic conditions is more likely to be of significance.

In this chapter we shall consider one area of the world where excessive utilization of irrigation has resulted in severe agricultural limitations because of the salinization of the soil. Salts build up to a level where food crops cannot survive. A paradoxical situation occurs in which the addition of irrigation water instead of helping to increase food production in some cases actually diminishes it. After considering the causes of salinization, we shall analyze the ways in which the situation can be remedied. The solution is not simple because researchers need to know all of the various pathways taken by the irrigation water. To understand how modern environmental scientists approach the problem we shall examine a computer model, which will eventually be used, that illustrates how a seemingly complex situation can be handled quite easily.

Archaeologists have long pondered over the fall of great civilizations. Their theories range from lead poisoning of the ruling classes, who used lead utensils, to destruction by marauding barbarians. Many historical accounts seem to indicate that a deteriorating physical environment was a contributing cause, and one highly intriguing theory suggests that the decline in efficiency of irrigation systems, on which a populace depended for its food supply, was responsible for the gradual disintegration of a society. The article, "Silt and Salt in Ancient Mesopotamian Agriculture", suggests that progressive soil salinization may have contributed toward agricultural decline, with all its potential ramifications, in parts of the ancient Tigris Valley [2:1251-58].

Salinization means that the salt content of the soil increases eventually to form a saline, nonproductive soil. It is a natural process that can occur in areas of average dryness where groundwater is available, but the process can become accentuated in arid and semiarid regions where irrigation water is added to the soil.

THE SALINIZATION PROCESS

Saline soils contain a high proportion of soluble salts including chlorides and sulfates of sodium, calcium, and magnesium. The accumulation of these salts in the soil is largely limited to arid and semiarid regions. In humid areas soluble salts in the soil are carried downward by the available water and transported via groundwater and streams to the oceans. This selective removal of materials from the soil, a process that is called leaching, cannot occur widely in dry regions. The lack of rainfall means that there is no water available to complete this flushing effect. The occasional exception does occur, however, and localized leaching produces water with a very high content of salt. This saline water tends to flow below the surface and accumulate in desert depressions where it causes the groundwater level--the water table--to rise. As the groundwater approaches the surface, it is drawn upward by capillary action in which water moves against gravity through the fine spaces between soil particles. Once at or close to the surface of the soil the water evaporates and leaves a concentration of salts in the root zone of plants or on the surface of the ground. Thus, the salinization process is a result of a combination of restricted, underground drainage, insufficient moisture for flushing salts from the soil, and a high groundwater level.

Salinity has an adverse effect on plants. It retards growth by limiting uptake of water. Often the salt constituents are toxic to plants. Boron, for example, which is frequently found in saline soils, proves toxic to many crop plants at concentrations of only a few parts per million. Efforts to develop salt-tolerant crops have been going on for many years, and although limited success has been achieved, the vast majority of basic food plants cannot survive in saline soils.

Given the normal conditions under which salinization occurs, it is evident that irrigation in dry regions can speed up the process. Suppose, for example, that irrigation water is added frequently and in relatively large amounts to an area where subsurface drainage is poor. There additions of water will lead to a rising water table, the beginning of salinization, and the lowering of crop yields. Consider, too, the role of irrigation ditches. Should these "leak," then the slow movement of underground water will again raise the water table (Figure 10-2). There is, clearly, an optimum amount of water and rate at which it should be added in an irrigation system. The rate and amount should reflect the water needs of the plants and recognize the amount of water passing down into the groundwater reservoir. Research in water supply, often using the concept of evapotranspiration--the transfer of water to the atmosphere through the combined processes of evaporation and transpiration--has resulted in establishing aids for computing the requirements of irrigation systems [6].

A MODERN CASE OF SALINIZATION

While salinization of the soil may have occurred in parts of ancient Mesopotamia, it certainly occurs in the world today. Pakistan is a notable example. "The twin menaces of waterlogging and salinity are posing a great threat to the agrarian economy in the Indus basin. Thousands of hectares of land go out of cultivation every year" [1:389] (1 hectare equals 2.47 acres). The effects in Sind, in the south part of the Indus Basin are especially severe (Figure 10-3).

For Pakistan the problem goes back to the latter part of the nineteenth century when the British set about irrigating Punjab and the Sind (Figure 10-3). By building dams called barrages they diverted water into great canals whose branches have a total length of tens of thousands of kilometers (one kilometer equals about .62 miles). The canals were carefully designed so that channel erosion and the deposition of silt were minimal, and, unlike the ancient inundation canals, they were able to carry water all year. Intensive agriculture, formerly confined to ribbons of land along river banks, spread to cover almost the entire region (Figure 10-4). The setting seemed right for extremely productive agriculture, but the physical conditions of the land raised unexpected problems.

Figure 10-2. **a.** Given sufficient inputs into a surface water system, water will pass to groundwater. In an irrigation system using canals, continued leakage causes the level of groundwater to rise. **b.** As it nears the surface, evaporation and transpiration of the water leaves a residue of salt and increasingly saline soils. (Redrawn from R.J. More, "Hydrologic Models and Geography," in R.J. Chorley and P. Haggett, eds., Physical and Information Systems in Geography, London: Methuen, 1967, p.177. (Used by permission.)

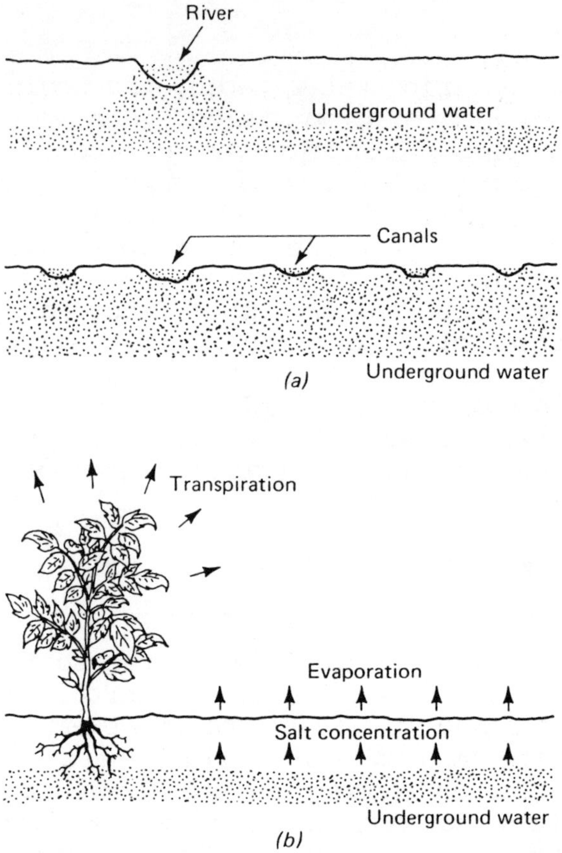

The average annual in-flow of water into the Indus and its tributaries is estimated at 170 million acre-feet, twice that of the River Nile. (Note: one acre-foot is equivalent to a layer about 0.4 hectare in area and about 0.3 meter in thickness). The flow is highly seasonal with almost half the volume being carried in July and August. During the time of high flow, the vast bulk of water flows unused to the sea. Some 84 million acre-feet are eventually added to the 9.3 million hectares (about 23 million acres) of irrigated land.

Figure 10-3. The lower sketch map illustrates the density of the canal system in Pakistan. (Note that one mile equals about 1.6 kilometers.)

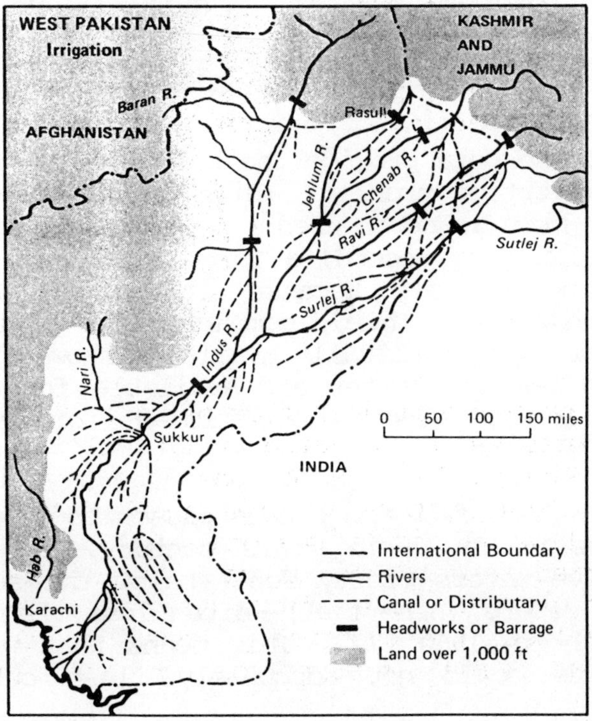

Figure 10-4. An irrigation ditch brings water to the dry lands of Pakistan. (Courtesy of United Nations.)

These data provide the basic reasons for the creation of the salinization problem. First, the seasonal flow of water means that much is never used for irrigation but flows directly to the Arabian Sea. Second, a large amount of the water in the canals is never used because it evaporates or leaks into the groundwater. Third, comparing the irrigation acreage with the water that is actually available shows that less than 0.6 meters (two feet) of water are available for each 0.4 hectare (about one acre), which is insufficient for crop production in such a dry climate.

Leakage from the canals passes to groundwater that has risen dramatically during this century, in some places from about twelve meters (forty feet) below the surface in 1905 to about 2.7 meters (nine feet)--often less than about 1.5 meters (five feet)--in 1960 [7]. This means that water added to irrigated fields could not adequately flush out the salts that were deposited in the upper soil levels. As a result, one of the main solutions to the salinity problem is to lower the water table. Accordingly, since the early 1960's, an organized effort has been directed toward that end, and declines of 0.6 meters (two feet) to more than three meters (10 feet) have been achieved [4: Q26-27]. The usual technique has been to install tubewells. In fact, the government of Pakistan reported in 1981 that under its programs "as many as 12,292 tubewells...[had] so far been installed... in Punjab, NWFP [Northwest Frontier Province] and Sind, involving more than 16.52 millions acres [c. 6.7 million hectares]" [5:197]. In addition 125,000 wells were estimated to have been installed in the private sector [1:486]. From tubewells saline water is pumped out and either mixed with canal water (to dilute the salt concentration) and then re-applied to fields or "exported" undiluted to salt lagoons, for example, in the Thar Desert. The withdrawn groundwater lowers the water table, and at the same time it supplies additional irrigation water to be applied to flush out the salts in the upper soil layers. Eventually, of course, the flushing effect will raise the salt accumulation in the groundwater again. To maintain a satisfactory "salt balance" some groundwater will need to be disposed of. Thus, in addition to exporting water to salt lagoons, it was suggested that over the next decade highly saline groundwater be pumped directly to rivers when they are in flood.

Such a generalized solution is easily visualized. There remained, however, certain important questions:

How should the tube wells be spaced in an area?
What should be the pumping rates of the wells?
How much of the pumped groundwater should be reapplied to the fields?
Do the canals need to be lined to stop water leakage?
What time period is involved in the effective lowering of the water table?
What crops are most applicable to the area being desalinized?
How much surface drainage is needed?

To obtain answers to so many variables it was necessary to construct a simulation model of the hydrologic processes involved. A schematic diagram of the model used is shown in Figure 10-5a. At first glance it looks enormously complicated, but analyzing its components makes it really quite easy to follow. Below the land surface there are two zones. In the lower zone, the groundwater zone, water occupies the air spaces present. The vadose zone is the subsurface area not filled with water. By simplifying the inputs and outputs (Figures 10-5b and 10-5c) of the system, the flows are more easily seen. The inputs are composed of river flow to canal (x), canal water that is applied to irrigation (u), and canal water that leaks from the canals (w). This input is augmented by rainfall (R), some of which goes to the canal (r") and some directly to groundwater Figure 10-5. Models of hydrologic processes showing inputs and outputs of water in a single well system. (r'). The outputs show that some rainfall (r" ') and some groundwater (S) pass out of the system by seepage to preexisting drains. Losses occur, too, through evaporation of groundwater (v) and through evapotranspiration of added irrigation water (u-u'). Finally, water pumped from the well (y) can either be recirculated on the irrigation area (z) or passed to river flow (y-z).

When one looks now at the complete diagram (Figure 10-5a) it is apparent that the symbol of a rectangle with a cross has been added at a number of places. These represent "valves" or decision points. At location 1, for example, how much water will be diverted from the river? At location 2, how much of the pumped water will be recirculated? Decisions made at such junctions determine the flow rates at any of the pump sites, and each is dependent upon local conditions. For each site an optimal solution incorporating agricultural production, engineering requirements, and costs can be deduced. Since the flow chart is easily transcribed into computer language, the optimum solution for each location is not difficult to ascertain. It still remains, however, for the computer output to be applied to the actual situation. Through a simulation model each of the questions listed earlier can be answered, but to put all the findings into the physical, economic, and social context of the real irrigation systems is a problem yet to be solved.

Figure 10-5. Models of hydrologic processes showing inputs and outputs of water in a single system.

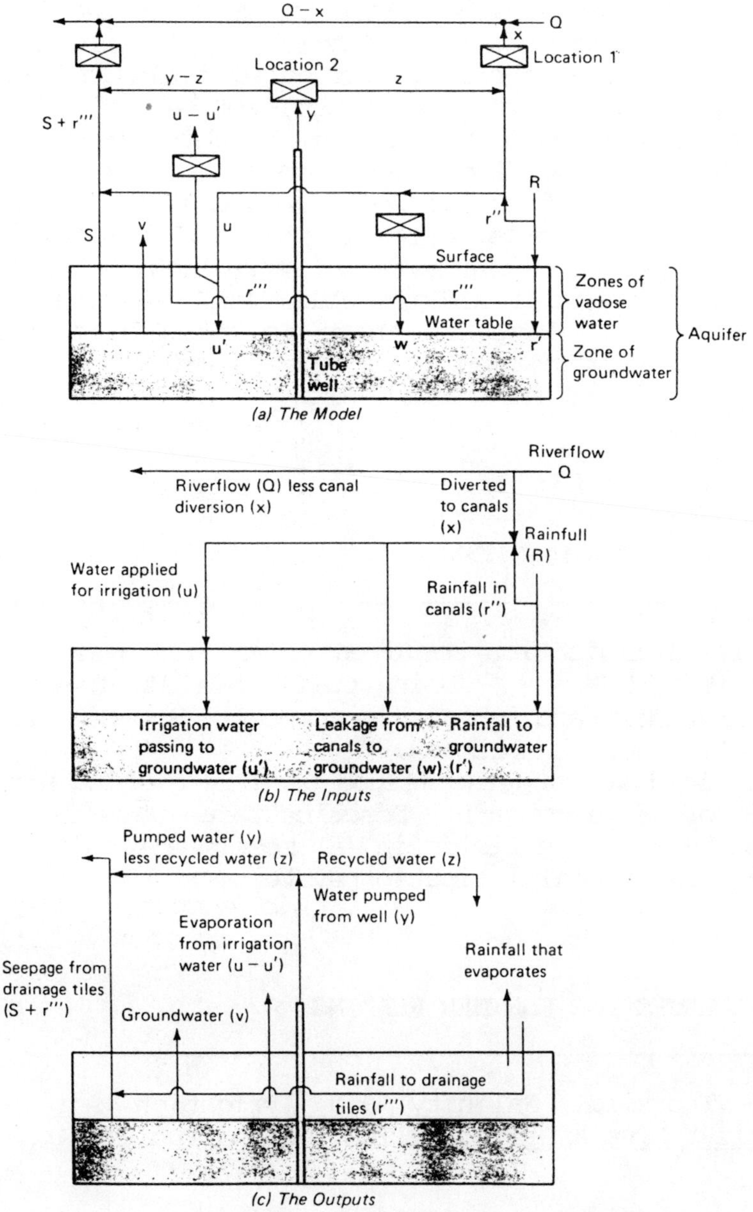

CONCLUSION

Although it deals with an old problem, the proposed solution to salinization in Pakistan incorporates modern analytic techniques. The proposed solution is based on the construction of a model, in this case a flow chart, applicable for computer use. It is a long way from an irrigated field in Pakistan to a sophisticated systems approach for collecting data. But such an analytic technique indicates the great advances that have been made in technical environmental studies. Pakistan's water problems have not, of course, been solved yet, but the type of research outlined here is certainly providing the means by which the physical problems can eventually be handled. With care, training, and financial aid the Pakistani salinization problem can be overcome. However, there is a human factor in the solving of environmental problems, and it is becoming increasingly evident that social change must accompany rational water planning.

SUGGESTIONS FOR FURTHER WORK

1. Extensive salinization could easily occur in irrigated areas of the United States. Using California as an example, show how the understanding of the concept of the water budget helps alleviate the problem.
2. To the field worker, the method used in getting water to the land is of prime concern. Describe three ways in which water is used to irrigate fields in (a) modern agricultural systems and (b) traditional irrigation systems.

SELECTED SOURCES FOR FURTHER READING

Jacobsen, Thorkild. Salinity and Irrigation Agriculture in Antiquity: Diyala Basin. Malibu: Undena Publications, 1982.

Schwab, G.O. et al. Soil and Water Conservation Engineering, Third Edition. New York: Wiley, 1981.
Sheridan, David. "The Underwatered West", Environment 23:2 (March, 1981), 6-13.
----------."The Desert Blooms--at a Price", Environment 23:3 (April, 1981), 6-24.
Thornthwaite, C.W., Associates. Average Climatic Water Balance Data of the Continents, Centerton, N.J.: C.W. Thornthwaite Associates, 1964, Vols. 16-17. (This group has published many volumes on water budget studies including continental surveys that provide data for many stations).
Turk, Jonathan and Amos Turk. Environmental Science. New York: Saunders, 1988, pp. 198-200, 211-221. (This is one of many introductory texts in geography containing accounts of irrigation systems).
U.S. Department of Agriculture. Water: The Yearbook of Agriculture. Washington: U.S. Government Printing Office, 1955.

REFERENCES

1. Akhtar, Rafique (Ed.). Pakistan Year Book, (10th Edition), Karachi: East and West Publishing Co., 1982.
2. Jacobsen, T. and R.M. Adams. "Salt and Silt in Ancient Mesopotamian Agriculture," Science 128, 1958.
3. Johnson, B.L.C. Pakistan, London: Heinemann, 1979.
4. Mundorff, M.J. et al. Hydrologic Evaluation of Salinity Control and Reclamation Projects in the Indus Plain, Pakistan, Water Supply Paper 1608-Q), Washington: U.S. Geological Survey, 1976.
5. Pakistan Ministry of Information and Broadcasting. Pakistan, An Offical Handbook, 1978-81. Islamabad: Ministry of Information and Broadcasting, 1981.

6. Thornthwaite, C.W. (Perhaps the best known method for computing water deficit is that introduced by American climatologist C.W. Thornthwaite. Details and examples of the methodologies involved are given in various publications). Centerton, N.J.: C.W. Thornthwaite Associates.

7. U.S. White House-Department of Interior Panel on Waterlogging and Salinity in West Pakistan. Report on Land and Water Development in the Indus Plain. Washington: U.S. Government Printing Office, 1964.

Chapter 11

Water in the Ground

The Context

In many parts of the world one of the most colorful local characters is the water witch, the person who uses a divining rod to locate underground water. These water diviners are not confined to remoter portions of the earth, and for most people it comes as a surprise that in the United States there are approximatedly 25,000 practicing water witches. The sheer availabilty of these specialists indicates the importance attached to finding usable groundwater in all parts of the country. In the United States groundwater is used as the main water supply system in both rural and urban area and its use often gives rise to rather special kinds of problems.

In this chapter we shall focus on two themes: one, groundwater related problems--overdrafting groundwater supplies where groundwater provides the main water supply; and two, land subsidence due to groundwater use characteristics.

GROUNDWATER DEPLETION IN THE GREAT PLAINS

With limited surface water supplies and related water deficits associated with high levels of evaporation in relation to precipitation, the primary source of water for irrigation in the High Plains of the United States is the Ogallala Aquifer; a massive underground water source 450,000 square kilometers (approximately 280,000 square miles) in area, extending through eight states (from South Dakota to Texas). The Aquifer supplies 30 percent of the groundwater used for irrigation in the United States. The groundwater system consists mainly of near-surface late Tertiary age sand and gravel deposits with saturated thickness levels ranging from 0-300 meters (0-1,000 feet). Almost half of the aquifer, however, has saturated thickness levels of less than 30 meters (about 100 feet) [5] (Figure 11-1).

The tremendous growth in land area irrigated and the associated demands on groundwater have had a profound impact on the attitude of resource managers and irrigators. As early as the 1930's, Webb [4] likened the Ogallala to a bank account of great size, maintained at a level by a balance of small deposits and small withdrawals. In essence, Western Kansas has been mining its groundwater resource, since water is being pumped from the Ogallala formation faster than the aquifer is recharging. Recharge, because of the region's low rainfall and depth to aquifer, is estimated to be approximately 6.4 millimeters (about .25 inch) annually. During most years, withdrawals are 10 to 20 times the recharge rate. At current rates of withdrawal relative to recharge, storage will be

Figure 11-1 Illustrates the variation of aquifer thickness within Kansas. (One meter is equivalent to about 3.28 feet).

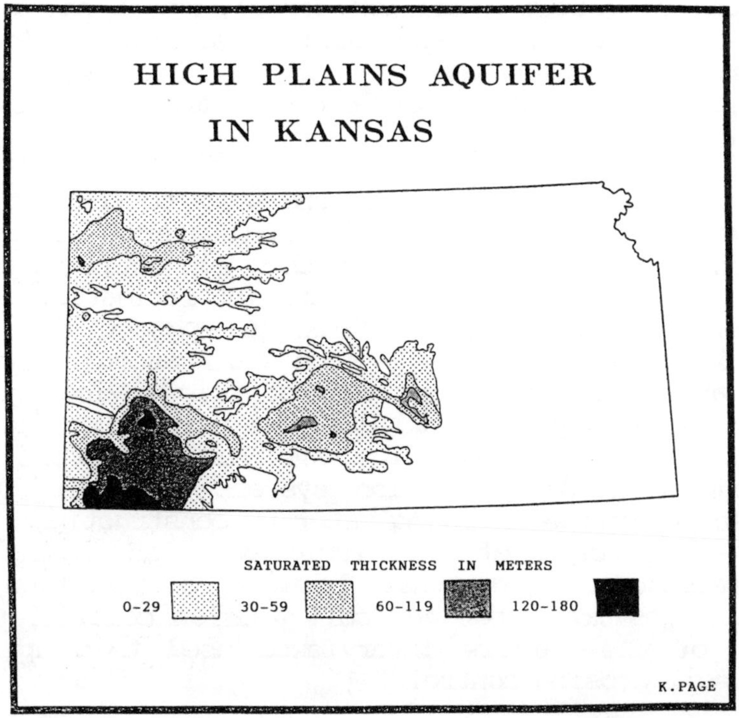

severely reduced in major portions of the Kansas High Plains by the year 2017.

In response to the concern for groundwater depletion in 1972, the Kansas legislature passed the Groundwater Management Districts Act. The legislature allowed for the formation of five major groundwater management districts. Since 1972, the districts have provided the basic framework for water resource planning in Kansas. To promote water conservation, the districts have implemented a variety of methods including limiting the number of new wells, controlling the spacing of irrigation wells, and providing conservation education.

These institutional forces have resulted in some changes in the rural system. Although irrigated acreage (primarily center pivot or circular systems) still amounts to approximately 800,000 hectares (about 2 million acres), there has been some shift from irrigated to dryland farming techniques. Additional adjustments include a switch to less water consumptive crops, improvements in irrigation efficiency, and improvements in general management skills.

When to apply water, how much, and when to shut it off are crucial elements to managing irrigation water. In the past, farmers turned the pumps on and let them run until they thought the field was well watered. Known as full irrigation, it often led to over-watering. Recent findings suggest that limited irrigation is producing nearly as much grain as full irrigation. Irrigators appear to be taking hold of this advice, given the response to the survey. Approximately 41 percent of the western Kansas irrigators surveyed in 1986 had converted to low pressure center pivot irrigation systems with drop nozzles. Twenty-three percent were saving water by constructing recycling pits, and 21 percent of the irrigators had reduced their watering frequency. In contrast, however, conservation tillage offers advantages for water and soil erosion control, yet only 46 percent of the farmers interviewed used this approach---mainly for soil erosion control [2].

Although the range of adjustments has had positive near-term results in slowing withdrawal rates, without massive water transfers into the High Plains, irrigation farming will end in most portions of the region in the next 50 to 80 years.

GROUNDWATER AND SINKING LAND

Although it is not as famous as the Leaning Tower of Pisa, the Basilica of Our Lady of Guadalupe in Mexico City has a similar problem. The towers on the west side of the Basilica are as plumb as the day they were constructed, but those on the east have a tilt that is easily discerned without surveying instruments. Subsidence of land beneath the east towers is responsible for the movement of the building.

In Venice, Italy, measurements of the Palazzo Loredan, the Townhall of Venice, show that its level above the sea has been decreasing since 1908. Observations such as this caused Carlo Berghinz to title a paper "Venice is Sinking into the Sea," [1] an account that anyone interested in preservation of the city's art heritage should ponder carefully.

By following a level line parallel to the Southern Pacific Railroad in the Elroy-Pichao area some 80 kilometers (50 miles) northwest of Tucson, Arizona, the United States Coast and Geodetic Survey has determined marked changes in the level of the land. Drawing upon precise leveling data obtained at various intervals since 1905, it was found that in the area the land has subsided as much as 1.1 meters (3.6 feet) [3].

These three different areas are all experiencing the same problem--subsidence of the land. Furthermore, the cause of the subsidence is the same in each case: the removal of water from below the ground. To fully understand this process, we will examine the situation in Mexico City.

THE AZTEC LEGACY: THE SITE OF MEXICO CITY

Mexico City is located in the Valley of Mexico, a basin some 2,280 meters (7,500 feet) above sea level. Surrounded by mountains, the closed basin receives runoff from the mountains through many small rivers, which until the fourteenth century flowed in to form a series of shallow lakes. It was on an island in the largest of these lakes, Lake Texcoco, that the Aztecs settled in 1325. By a system of drainage canals, dikes, and causeways, the shallow lake was drained and its former bed became the site of the center of the Aztec Empire and, eventually, the site of Mexico City.

The rock strata below the old lake bed consist largely of debris deposited by streams that flowed into the valley in

Pliocene and Pleistocene times. Consisting of sands and gravels with some interbedded silts, these strata form an ideal aquifer, or rock formation through which water readily flows. They are overlain by volcanic ash, fine-grained lake deposits, and sediments recently deposited by water. The location of the aquifer in relation to the surrounding mountains gives rise to a natural, artesian groundwater system. Figure 11-2 illustrates the requirements for such a system. When modern Mexico City grew on the site selected by Aztecs, this groundwater supply seemed to provide the ideal answer for its water needs. In fact, until the 1960's most of the municipal water supply was derived from wells within the city.

It was subsidence of the earth surface that caused the use of the artesian well system to be questioned. From observations taken between 1898 and 1956, it was estimated that total land subsidence in the Mexico City area varies from 5-7 meters (16-23 feet) [5]. What appears to be happening is that withdrawal of water from the aquifer reduces the artesian pressure (Figure 11-2). With no water in the tiny spaces between individual rock grains, the ground becomes compacted. Compaction of material means that it occupies less space, and, as a result the surface above the compacted strata subsides.

In a city the size of Mexico City land subsidence causes many problems. Some of the problems are obvious, such as the tilting of buildings. Others are less easily visualized, such as the problems of sewage disposal. In Mexico City subsidence is so great that sewer lines, instead of flowing downward away from the city, now run uphill so that normal gravity flow is in the wrong direction. As one government official noted, if the pumps ever failed, sewage would back up and flow into many parts of the city. To help overcome this problem, tunnels are being constructed to help drain the existing sewer system.

It is evident that officials are well aware of the subsidence problems in Mexico City. To prevent the decrease of pressure in the artesian system, it is necessary to reduce the use of groundwater. To this end comprehensive plans have been made, and are being instituted, to bring water into the city from newly constructed reservoirs. At the same time, recharge wells have been constructed to inject floodwaters into the aquifer.

Figure 11-2. A cross section of strata illustrates the artesian pressure system.

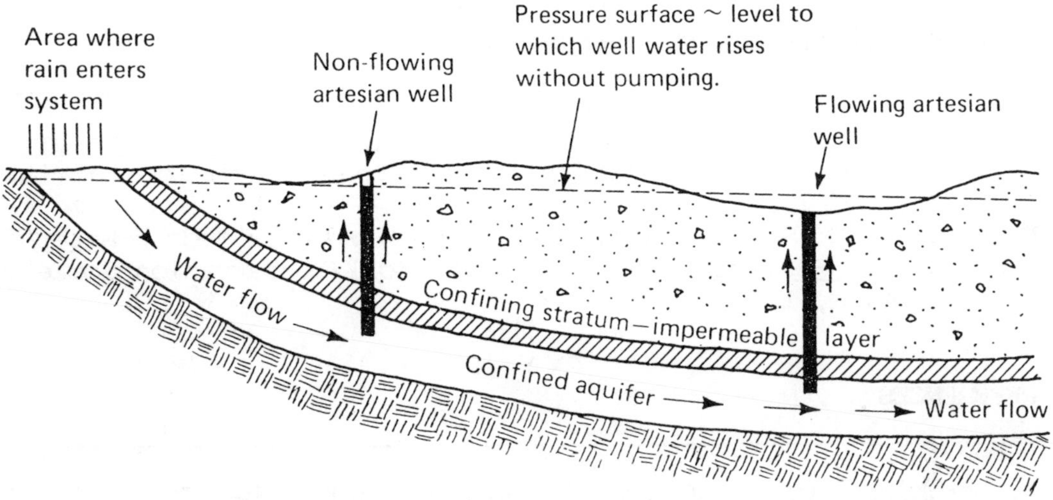

CONCLUSION

From these examples we have some indication of the extent of the problems associated with the use of groundwater resources. The examples also illustrate how the use of groundwater results in both immediate and long-range planning problems. However, when the problems of depletion and subsidence are looked at in a detached manner, the solution is really quite simple: stop withdrawing groundwater and institute reinjection programs. Such a recommendation is, of course, quite unworkable, given present needs. Like most environmental problems, the eventual solution rests not only with researchers but with economists, politicians, and the public.

SUGGESTIONS FOR FURTHER WORK

1. Are water transfers a feasible solution to addressing water depletion problems in the Ogallala Aquifer region?
2. Land subsidence is a major problem in parts of Florida. Is the process of subsidence similar to those discussed in this chapter?

SELECTED SOURCES FOR FURTHER READING

Henderson, T.R., J. Trauberman, and T. Gallagher. Groundwater: Strategies for State Action, Washington: Environmental Law Institute, 1984.

Napier, T., D. Scott, K.W. Easter, and R. Supalla, eds. Water Resources Research, Ankeny, Iowa: Soil Conservation Society of America, 1983.

REFERENCES

1. Berghinz, C. "Venice is Sinking into the Sea," Civil Engineering 41 (1971), pp. 67-71.
2. Nellis, M.D., "Estimating Water Demand in Northwestern Kansas," Manhattan, Kansas: Kansas Water Resources Research Institute, (1987), 36 pp.
3. Pland, J.F. and G.H. Davis, "Land Subsidence Due to Withdrawal of Fluids," Reviews in Engineering Geology, 2 (1969), pp. 187-269.
4. Webb, W., The Great Plains, Boston: Ginn and Company, 1931.
5. Weeks, J. and E. Gutentag, "The High Plains Regional Aquifer - Geohydrology," in: Proceedings of the Ogallala Aquifer Symposium II, Whetstone, G., ed., Lubbock, TX: Texas Tech University Water Resources Center, (1984), pp. 6-25.

Chapter

12

Problems on the Beach

The Context

About 71 percent of the earth's surface is water. Land, ranging in size from huge continental masses to the smallest atoll, is scattered across the face of the earth, creating thousands of miles of coast. In these coastal regions changes are always taking place. In some cases, such as when an island made of unconsolidated material vanishes after a great storm, the rate of change is extremely rapid. In other instances, no apparent shoreline change can be observed for decades.

Whatever the rates of change, however, the effects are inevitably felt by human beings.

Coastal zones have been particularly important to human beings throughout history. Tiny fishing villages and large ports both illustrate the close relationship between land and the sea. It is little wonder that the study of coastlines is an important part of geography. In this chapter we shall look at two kinds of coastal change. The first concerns problems associated with coastal erosion and deposition. The second considers changes resulting from pollution associated with oil spills. Each points to our vulnerability in changing or maintaining coastal areas.

THE CHANGING COASTLINE

The coastal zones of the world are fascinating places. Along these narrow strips of land the energy of the sea is constantly being dissipated. Land is eroded and material is transported, eventually to be deposited elsewhere. The study of coastlines is a study of change, and intimately connected with that change are the small efforts of people to ward off the inexorable work of the sea. The battle with the sea has gone on for hundreds of years, but with the growth of population and excessive use of the coastal areas, the effects of natural changes have frequently become secondary to changes that result from our attempts to modify the natural coastal system. The situations we shall explore in this section show how the best plans often do not work. In one case the construction of artificial harbors is the problem, and in the other, the failure to appreciate the relative rapidity of coastal erosion has caused a different problem.

SANTA BARBARA AND SANTA MONICA

One of the joys of visiting a beach is simply watching the waves, but wave watching can tell us a lot about the relationship between the land and the sea. If you observe carefully, you will note that the advancing wave line is not always parallel to the beach and that water from the breaking wave flows up the beach in the direction of the wave. In receding, however, the water flows straight down the beach to the sea. If a paper cup or old sneaker is in the wave zone, you will see that it follows a path that causes it to migrate along the beach. You have just observed longshore and beach drift.

It is this very same drift that has created problems along the coast of California just north of Los Angeles. At that location the coast changes from a predominantly north-south to a predominantly east-west orientation. Waves, which come mostly from the west, produce a longshore drift of some 200,000 cubic meters (260,000 cubic yards) of sand each year along the coast south of the point.

At Santa Barbara a sandy beach of variable width used to be open to the Pacific Ocean. In the late nineteenth century the inhabitants began to promote construction of a harbor that could handle deep-water ships. Between 1873 and 1921, the United States Army Corps of Engineers prepared a number of reports on the feasibility of the project. All were unfavorable. Nonetheless, with local funds an L-shaped breakwater some 450 meters (1,500 feet) long was completed in 1929 (see Figure 12-1a). Initially, the breakwater did not extend to the shore on the west side of the structure, and sand poured through the gap between the breakwater and the coast to settle in the harbor. To stop the harbor silting, the west side of the breakwater was extended almost to the shore.

However, the sea continued to transport debris along the coast. Just seven years after the construction of the breakwater, deposits behind the shore-connected pier amounted to almost 1.5 million cubic meters (2 million cubic yards). As the area filled, the sand slowly accumulated along the outer edge of the breakwater, rounded it, and was deposited inside the harbor. The harbor became so filled with sand that it could be used only by shallow draft boats, but this silting represented only part of the problem. Under natural conditions the sand would pass down the coast to replenish the beach sand. The lack of this

Figure 12-1. These two diagrams show how longshore drifting is modified by the construction of breakwaters at (a) Santa Barbara and (b) Santa Monica.

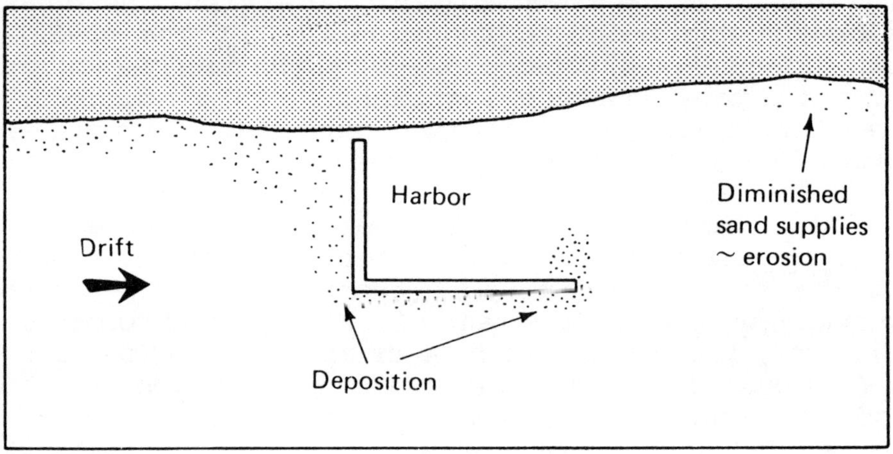

(a)

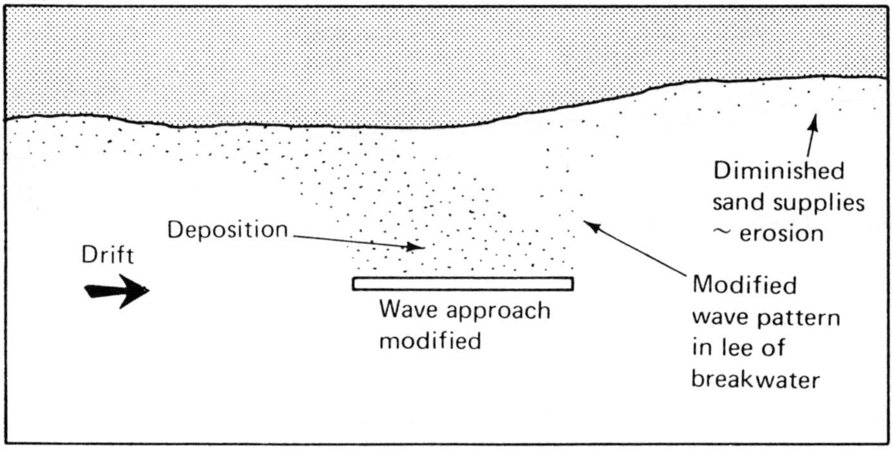

(b)

replacement sand meant that serious erosion occurred to the east of the breakwater. At Miramar Beach, 6.5 kilometers (4 miles) east of the harbor, a wide sand beach was largely destroyed. Just 16 kilometers (10 miles) east, removal of the beach led to the undermining of cottages, and between 1929 and 1940 the beach at this point was cut back some 72 meters (240 feet).

A number of solutions to the problems were attempted. Francis Shepard and Harold Wanless report that a government dredge removed some 152,000 cubic meters (200,000 cubic yards) of deposits from the Santa Barbara Harbor and redeposited it in the water about 1.7 kilometers (approximately one mile) east of the harbor. It was anticipated that the relocated sand would be moved by waves along the coast to the area suffering beach loss. This redistribution did not happen, and the sand remained where it had been dumped, while erosion of the beaches continued to create problems. Shepard and Wanless further describe the situation.

> Following these catastrophic results of the construction of the Santa Barbara Harbor, the Army Corps of Engineers investigated the whole problem of this beach erosion and harbor shoaling and proposed several solutions. As a result, in June, 1938, a hydraulic pipeline-dredge dumped about 600,000 cubic yards [450,000 cubic meters] of sand from shoals in the harbor to a shrunken beach about one mile [1.6 kilometers] east of the Santa Barbara pier. Consequently, all of the beaches for many miles to the west have been partially replenished. Dredging has been repeated at 2- or 3-year intervals, and the process has been reasonably satisfactory. The cost has exceeded $30,000 per year and during the intervals between dredgings the harbor has been extensively filled with sand. The authors visited Sandyland in June, 1968, and found that the shore is now protected by a rubble seawall. Nearer the harbor, at Miramar, a narrow beach has developed, and the cottages are apparently no longer threatened [5:291].

The residents of Santa Monica also wanted a sheltered harbor. They planned to construct a breakwater parallel to the shore in order to avoid the problems created at Santa Barbara. It was assumed that the sand moving eastward along the coast would not be hindered by such a breakwater (see Figure 12-1b). The breakwater was constructed in 1933, but it did not work as they had anticipated. Wave movement in the lee of the breakwater differed from that in the unprotected areas. An area protected from the full impact of the incoming waves--a "wave shadow"--was formed by the interruption of the natural southeast movement of the waves toward the shore. The beach behind the breakwater edged outward and eventually became a trap for eastward moving sand, while the inevitable erosion of the coastline occurred downshore. Like Santa Barbara, the Santa Monica Harbor now employs a dredge.

VANISHING COASTS

The two cases cited above concern the movement, or non-movement, of sand as the basic shoreline problem. Settlements on top of cliffs are also faced with problems. Cliffs are continually subjected to the erosive power of the waves, and buildings near the edge of cliffs are particularly vulnerable. The struggle to overcome coastal erosion is remarkably well illustrated on the east coast of England where cliff erosion and beach formation have been recorded for centuries. The historical evidence provides some fascinating stories.

On the coast of Lincolnshire (Figure 12-2) the town of Mabelthorpe is located on relatively soft glacial deposits that form low cliffs. Since the thirteenth century the sea has been making inroads into the cliffs, but it is estimated that in the last 400 years the sea has advanced as much as eight kilometers (five miles) over wide areas. The problem, however, started many years before. Chronicles tell us that St. Peter's Church in Mabelthorpe was "rent asunder" by the sea in 1287. The problem became acute in 1443 when it was recorded that the lord of the Mabelthrope Manor was exempted from paying taxes because

Figure 12-2. The coastal area between Flamborough Head and the Wash on the east coast of England.

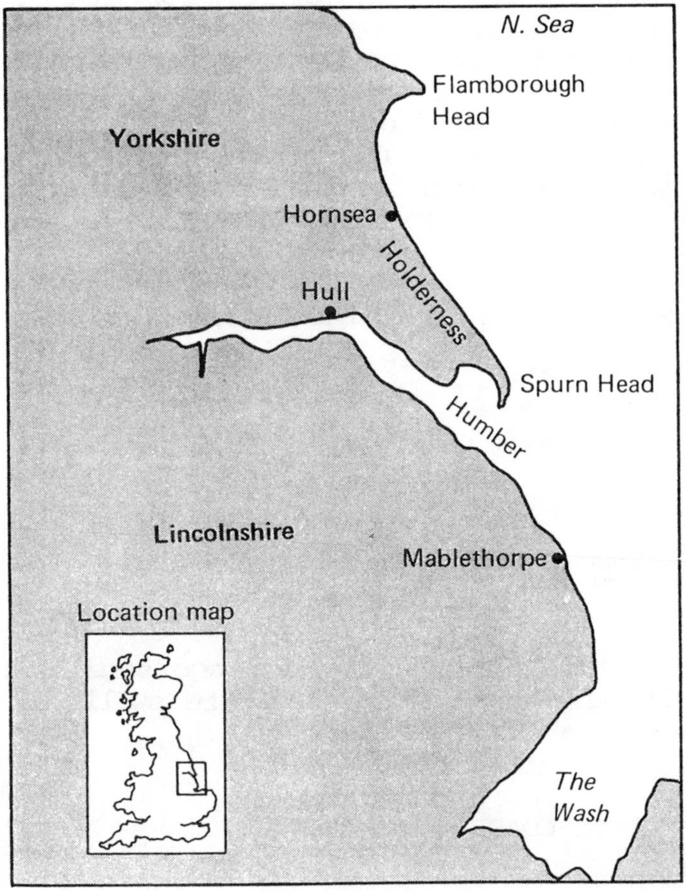

his money was needed to make defenses against the invasion of the land by the sea. Eventually the sea won, for about 1540 St. Peter's was destroyed and "both church and chancel were swallowed up with the sea" [2].

Evidence such as the disappearance of Dunwich Church has been used to estimate the rate of erosion in this part of the world.

At Holderness, just north of Mabelthorpe, the cliffs are between three and nine meters (10 to 30 feet) high (3 to 9 meters) and are made of easily eroded glacial drift. Only narrow sandy beaches run along the foot of the cliffs, for most of the eroded material is swept southwards by the prevailing waves. Little material is brought in from the north where the chalk headland, called Flamborough Head, limits southward passage of drifting material. The soft cliff deposits and the lack of protective beach lead to rapid erosion. One estimate of erosion along this coast is by Valentin, who notes that slight changes in the sea level modify the rate of change [6:296-315].

Rates of Erosion along the Lincolnshire Coast

1852-1889	Low Sea Level	1.52 m/yr	(5.02 ft/yr)
1889-1908	Rising Sea Level	1.75 m/yr	(5.81 ft/yr)
1908-1926	High Sea Level	1.88 m/yr	(6.20 ft/yr)
1926-1952	Slowly Falling Sea Level	1.72 m/yr	(5.61 ft/yr)

At this rate a house built 100 years ago some 17 meters (55 feet) from the cliff edge would today be swallowed up by the sea.

MEASURES TO PREVENT COASTAL EROSION

As we have seen, coastal erosion in the United States is a major problem. Most beaches of North Carolina, for example, are retreating at a rate of 0.9 to 1.8 meters (3 to 6 feet) a year. Eighty-six percent of California's coastline is eroding [4].

A number of devices such as seawalls, riprap walls and groins have been constructed in an attempt to halt or reduce beach erosion, but with mixed success. Groins of rock or concrete, built perpendicular to the shoreline, have been constructed to slow the littoral draft, thus reducing the effectiveness of the current to transport the sand.

The United States Coastal Zone Management Acts of 1972 and 1980 provided federal funds to 37 coastal and Great Lakes states for assisting them in the development of programs for protecting and managing coastlines. By early 1987 more than 90 percent of the eligible states had developed coastal management plans.

OIL ON WATER

As though problems of erosion and deposition were not enough, the residents of coastal areas now face another problem, which is a result of the ever growing needs of our technological society. With the demands for oil that exist today, drilling teams have moved from the land to offshore areas, and along the coastline drilling platforms are familiar sights. To transport oil, tankers with capacities of 180,000 metric tons (200,000 U.S. tons) have been constructed. Technology isn't all perfect; oil spills from accidents either in the offshore drilling or on the supertankers add to the problems of the shores. An oil spill at Santa Barbara is a classic case of oil-related coastal problems and serves as an apt example.

At an oil platform located some nine kilometers (about 5.5 miles) south of Santa Barbara, a well was drilled to a depth of 1,060 meters (3,479 feet). Casing for the well extended to a depth of 157 meters (514 feet), about 65 meters (240 feet) below the ocean floor. On January 28, 1969 the drill pipe was being removed when mud began to flow from the hole. The drill team dropped the drill pipe into the hole, and the flow of mud ceased. Soon after, however, gas and oil began to boil up through the water near the platform. With one exception, located some six meters (20 feet) from the platform, most of the boils subsided the next day. However, the main blowout continued to discharge large volumes of oil into the sea. It seems that "a small fault outcrops on the ocean floor near the platform. The location of the outcrop and strike of the fault appears to coincide with the location of all the boils which have been observed. Union Oil Company expressed the probability that the gas and oil escaping from the ocean floor were flooding up through the hole, laterally through porous strata, and then to the ocean floor through the small shallow fault" [1:15-27].

Attempts to stem the flow of oil continued. Additional drill pipe was added and mud was pumped into the well, but to no avail. A relief well was started as a backup measure while seawater, weighted with calcium chloride and mud, was injected into the bore hole. Finally, on February 7, it was decided to plug the hole with cement. A day later, the president of Union Oil Company announced that the well had been plugged. The announcement was premature, for since that time, leakage of oil and gas, although at a reduced rate, it was estimated that more than 12,000 kiloliters (3.25 million gallons) of oil had spilled into the channel.

In another part of the world a large amount of oil was released when the oil tanker Torrey Canyon was stranded on the Seven Stone Reef off Lands End in Cornwall, England, in March, 1967. Ten days after the ship had run aground her entire cargo, some 107,000 metric tons (118,000 U.S. tons) of Kuwait crude oil, had escaped from her tanks. When the ship foundered, the original hope was that it might be refloated or that the oil could be transferred to another vessel. Such a hope delayed government action for 10 days. Then, after the pounding waves had broken her back, aerial bombing was ordered to burn the remaining oil. Prior to the bombing, oil drifted off in three main masses. Some 27,000 metric tons (30,000 U.S. tons) passed up the English Channel toward the French coastline; 18,000 metric tons (20,000 U.S. tons) floated to West Cornwall; and 45,000 metric tons (50,000 U.S. tons) found their way south to the Bay of Biscay.

On March 24, 1989, the oil tanker, Exxon Valdez, ran aground on Bligh Reef in Prince William Sound off the coast of Alaska. Shortly after, approximately 38,000 kiloliters (10 million gallons) of crude oil gushed from the tanker's lacerated tanks into one of the United States' most sensitive environments. Because the Sound is protected from the open sea, the slick was slow to dissipate. Toxic hydrocarbons permeated water and the shorelines, continuously seeping into the food chain. The final toll has yet to be determined, but approximately two-thirds of the diving birds have disappeared. It has been estimated that damage claims would cost Exxon at least $500 million [3]. That would represent less than 7 percent of Exxon's profit for 1988.

The costs to Prince William Sound fishermen were numerically smaller, but proportionately much larger. In 1988, those waters yielded $131 million worth of salmon, herring, halibut, and shellfish. The spill occurred just weeks before the herring were to begin spring runs, and before the scheduled release of millions of salmon fry expected to return to the Sound as marketable fish in 1991.

Another impact occurred to the area's 5,000 sea otters. They are especially vulnerable to oil because it reduces the insulation and buoyancy of their fur. If they do not die from cold, they can ingest fatal doses of oil, while licking it off their coats. Of the dozens of otters brought in for cleaning, only a few could be saved. Hundreds more died slow deaths on remote, oil-fouled shores.

Other long term and less visible effects remain to be catalogued: on the plankton that are the foundation of the food chain in the Sound; on the spawning habits of herring and salmon that survive the spill; on bald eagles that are dying; on bears that may die from eating polluted animals; and on whales that pass through the area. The list is long and the prospects gloomy.

Although these spills have generated major environmental consequences, the good news is that the estimated input of oil into the ocean has decreased in the past decade by fifty percent [4]. Improved safety measures, including better navigation equipment, training, control and tougher safety standards for tankers, have had positive effects.

The inputs from the spills that do occur, however, are significant. Hydrocarbons kill large numbers of aquatic organisms. Floating oil can coat feathers of birds, preventing natural buoyancy, and most drown. It is estimated that 150,000 to 450,000 marine birds are killed each year by oil pollution in the North Sea and North Atlantic regions [4].

CONCLUSION

Environmental problems associated with the use of coastal zones are currently receiving considerable attention from both federal and private research agencies. In some cases the focus is on the preservation of existing conditions; in others, researchers are seeking designs that will have minimal environmental impact and yet will maintain safe conditions for residents of the shore areas. Of course, offshore oil resources are of prime importance in the current national energy crisis. It is to be hoped, however, that economic considerations do not totally outweigh potential ecological problems.

SUGGESTIONS FOR FURTHER WORK

1. One of the most remarkable coastline changes is the reclamation of land in the Netherlands. Explore the way in which the Dutch implement their schemes and outline the setbacks that have occurred.
2. The call for conservation of coastal wetlands has mounted in recent years. Why should wetlands be preserved?

SELECTED SOURCES FOR FURTHER READING

Bird, E.C.F. Coastline Changes: A Global Review, New York: John Wiley and Sons, 1985.

Buther, M. "Netherlands: Dutch Continue to Reclaim Land from the Sea," Science, Vol. 176 (1972).

Kaufman, W. and O. Pikley. The Beaches are Moving: The Drowning of America's Shoreline, Garden City, New York: Anchor Press/Doubleday, 1979.

REFERENCES

1. Holms, R.W. "The Santa Barbara Oil Spill", in Hoult, D.P., ed., Oil on the Sea, New York: Plenum, 1969.
2. King, C.A.M. Beaches and Coasts, 2nd Edition, London: Edward Arnold, 1972.
3. Lewis, T. "Tragedy in Alaska", National Wildlife, Vol. 27, (1989), No. 4, pp. 4-9.
4. Miller, G.T. Living in the Environment, Belmont, California: Wadsworth Publishing Co., 1988.
5. Shepard, F.P. and H.R. Wanless. Our Changing Coastlines, New York: McGraw-Hill, 1971.
6. Valentin, H. "Der Landverlust in Holderness, Ostengland von 1852 bis 1952", Die Erde, Vol. 6, (1954). (King, Beaches and Coasts (listed above) provides a synopsis of this study).

Part IV
The Geosphere

Chapter 13

Continuing the Work of Wegener

The Context

There are few geological theories as well-known as continental drift. Most people have never herd of Bowen's reaction series and geosynclinal theory, but they have probably heard of the continental drift. Perhaps it is the grandeur of the idea that catches people's attention. Whatever the reason, it is a fascinating concept. The person who deserves the most credit for developing the concept is Alfred Wegener. His work met with a great deal of criticism, but from it, along with the contributions of many other scholars, came an understanding of the evolution of the earth's crust.

Had not great strides taken place in the geosciences, the idea of continental drift might have stagnated. Fortunately, new geophysical methods were found and new light was thrown on Wegener's idea. Today, as in the past, the creation of major geologic division--the continents and the oceans--is again in the forefront of research and, besides providing fascinating data, will hopefully help explain such violent phenomena as volcanoes and earthquakes.

The development of natural and physical sciences in the nineteenth century required bold, farseeing individuals to formulate hypotheses and theories to account for the worldwide observations that were being accumulated. The nineteenth century certainly did produce such scientists, who, like Charles Darwin, proposed theories that pointed to the origin of life and its relation to the environment. However, such thinking was not limited to the nineteenth-century naturalists. A name that might well be added to the list of men who became synonymous with the development of natural science is Alfred Wegener. Born in Germany in 1880, Wegener made a great contribution to earth science, beginning in 1906 and ending with his untimely death in 1930.

Wegener trained as a meteorologist, and, as such, he joined four expeditions to Greenland. It was on his final expedition, in November, 1930, that he left a base in central Greenland and never returned. However, Wegener is not best known for his work in meteorology. Rather, his name is linked with his great work, The Origin of Continents and Oceans, in which he set forth what is known today as the theory of continental drift. [2] He thought that the continents were once united and have since broken up into small and large pieces which are moving continually apart.

The idea of the movement of continents was not original with Wegener. The nineteenth-century geologist, Eduard Suess, had identified a super continent that he had named Gondwanaland. It was Wegener, however, who put the idea into a scientific framework. The evidence that he presented falls essentially into four main categories. Perhaps his best known idea is the jigsaw puzzle relationship between the continental outlines. Wegener wrote: "The first notion of the displacement of continents came to me...when, on studying the map of the world, I was impressed by the congruency of both sides of the Atlantic coasts, but I disregarded it at the time because I did not consider it probable." [2] To make it more probable, Wegener studied the paleontological, or fossil, evidence of former land connecting Brazil and Africa and analyzed the results of earlier research in the areas of geology and paleontology.

The research proved worthwhile, and he found a second piece of evidence. Fossil plants and animals, believed incapable of crossing deep water, appeared in limited areas on both sides of the Atlantic. This evidence itself gave rise to heated debates. Could it not be that, at some time in the past, South America and Africa were joined by a land bridge? Such an explanation would certainly account for the distribution of fossils, and the legendary continent of Atlantis could have perhaps been the land bridge.

A third area of evidence offered by Wegener was the structure and makeup of rocks on either side of the Atlantic. By tracing mountain chains created at various geologic times, he found that equivalent formations occurred on either side of the Atlantic Ocean. As Figure 13-1 illustrates, the mountains of Scotland and Scandinavia appear once to have been continuous with mountains of the same geologic age in North America. Similarly, the great mountain-building forces that produced the Alps and other high mountains of Europe and North Africa can be related to patterns of similarly aged mountains in the Americas. Finally, remnants of mountain systems in South America appear to be related to those in South Africa. This evidence was not

Figure 13-1. Mountain systems of similar ages (crosses, circles, or dots) occur on either side of the Atlantic.

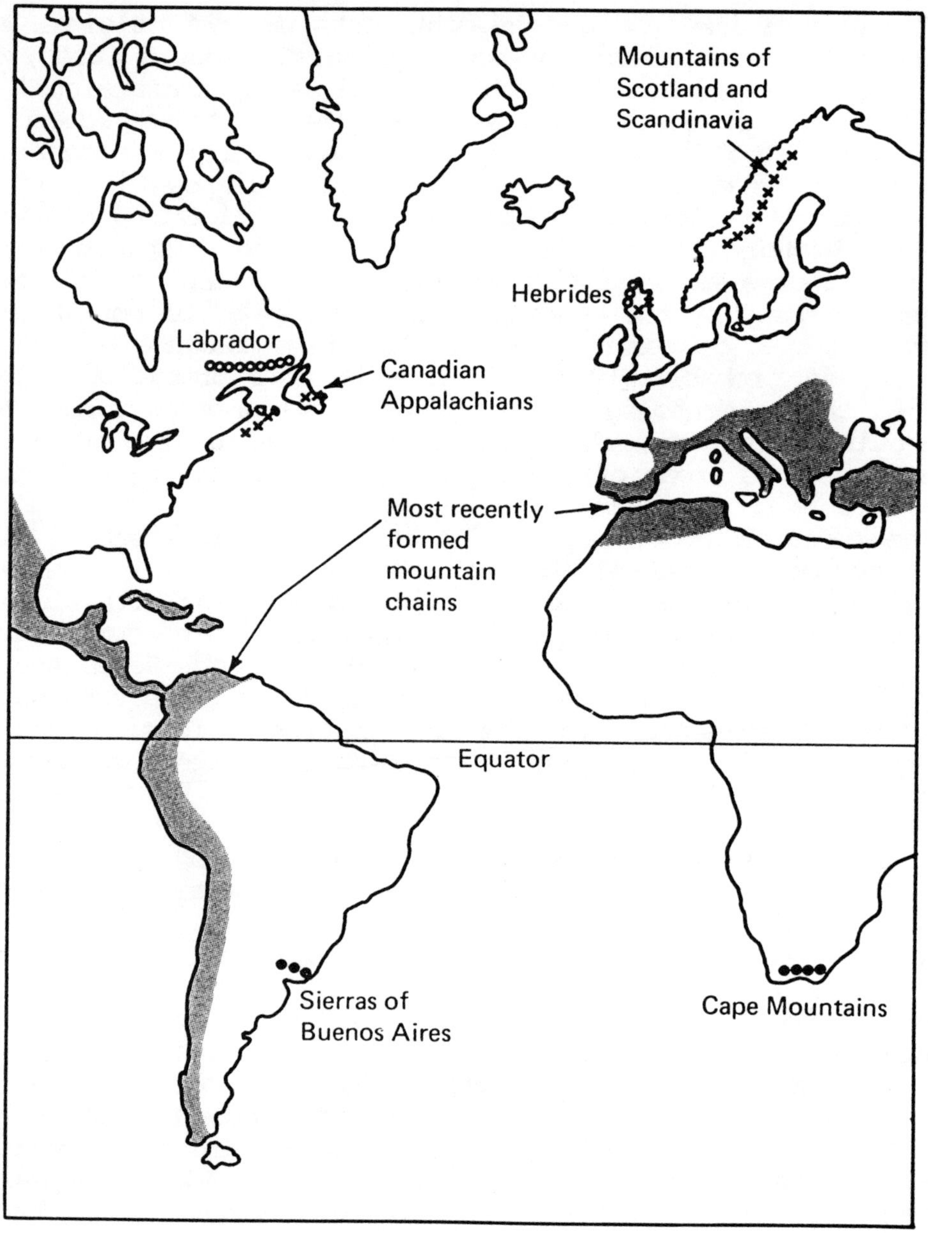

readily accepted by all geologists, and arguments against it were strongly voiced. Some critics suggested that while a similarity did exist on either side of the Atlantic, it was far from a perfect match, and the differences were significant enough to negate the idea that the continents were once joined.

However, Wegener's idea did generate a lot of respect when evidence derived from ancient climates was related to his theory. One of the periodic ice ages that has occurred on earth took place some 300 million years ago. The extent of this glacial activity can be estimated by mapping the evidence of ice erosion and deposition. Most traces of this ancient glaciation are confined to the southern hemisphere with a major ice-formation center in what is now South Africa. Traces of the work of ice in South America, in which no ice-formation center occurred, are best explained by assuming that the glaciers actually originated from the ice source in South Africa.

There remained one big argument that Wegener and his supporters could not answer. What was the enormous driving force that pushed the continents apart? What is the mechanism of continental drift? The major force that Wegener identified was not, unfortunately, a viable one. He suggested that a "Polflucht Kraft", a pole-fleeing force, was at work. This force resulted from a difference between the center of gravity of a continental mass and--assuming that a continent floats on denser rocks--its center of buoyancy. If these two do not coincide, the force that results between them will be directed toward the equator. Calculations have shown, however, that the feeble force generated by this difference is not sufficient to make the continents move.

Other possible causes of continental drift were introduced by scientists who believed in the basic correctness of Wegener's theory. One of the most important was that introduced by geologist Arthur Holmes. The cause, he suggested, was not gravity or buoyancy but currents created by differential heating beneath the earth's crust. This idea put a new perspective on continental drift. Rather than the continents being the active elements, Holmes considered them passive, carried along on convective currents like objects on a conveyer belt. Figure 13-

2 illustrates the idea. The heating of materials deep within the earth's mantle creates an ordinary convective current in which warmer material rises toward the surface, flows outward, and eventually descends in cooler areas (a). At the areas where these convection currents rise, the crustal portions are dragged apart (b). Such an explanation offered an alternative to the

Figure 13-2. Holmes's concept of drifting continents based on subcrustal convection currents in the earth mantle.

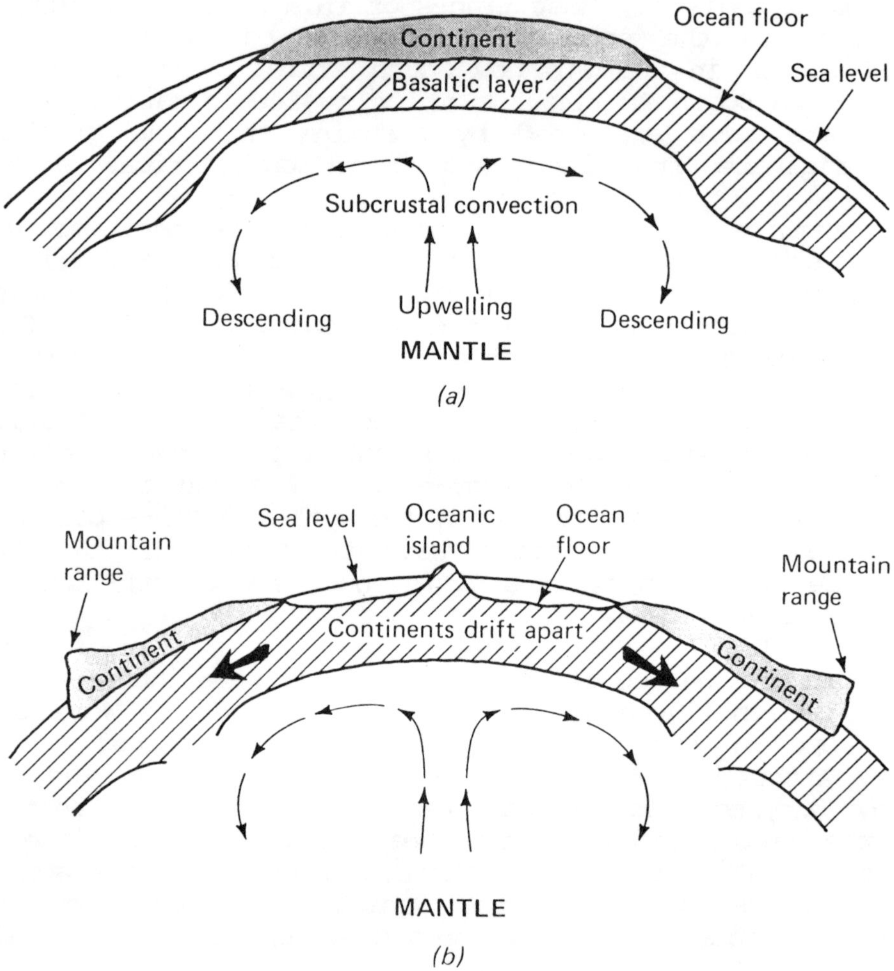

inadequate pole-fleeing force, but many facts about the source of the differential heating and the reasons for the location of the currents were not easily explained.

By about 1940, scientists had explored all of the ideas that were available to them, and the debate stagnated. It was through exploration of the ocean floor and the development of new geophysical techniques that new light was cast upon the problem. The whole concept of moving continents was revitalized, and a revolution occurred in the earth sciences.

THE NEW LOOK

The way in which continental drift is viewed today differs from the basic ideas of Wegener in a number of ways. This new look draws upon data and techniques that were unavailable to him, and we must understand the theories and research that made the advances possible. Many of the insights came from the study of the ocean floor. It has long been known that the ocean floor is composed of great basins interrupted by mountains that rise to--or almost to--the ocean surface. In some cases the mountains are isolated features, but frequently they form great chains stretching over thousands of miles. Studies of these mountain chains showed some surprising irregularities. By 1960 marine geologists from Columbia University had found that the temperature of the ocean floor near the ridges, as the undersea mountain chains are called, was higher than elsewhere. Further, the amount of sediment on the ocean floor near the ridges was much less than had been thought. The lack of sediment was surprising because the ocean floors were assumed to be very old and should, therefore, have been covered by layers of sediment. As it was, the depth represented deposits of only a few hundred million years.

To account of these observations, the late Professor Harry Hess proposed a model that refined Holmes's convective theory of continental drift and, at the same time, drew up a different representation of the concentric layers that make up the earth.

Formerly, the continents had been considered as light rock masses resting upon denser rocks, which were comprised of material similar to the rock called basalt. This rock, in turn, rested on what was called the mantle of earth. According to the newer interpretation, the crust is made up of both continents and ocean floors that rest upon a rigid lithosphere. Below the lithosphere is a partially molten layer called the asthenosphere. Heat derived from radioactive elements creates subcrustal convection currents in the asthenosphere. These currents cause movement in the lithosphere, and the continents move as well. Hess further suggested that where rising and spreading currents occurred, the lithosphere, or plate, would be dragged apart laterally, and volcanic activity would provide the material to make new crustal rocks. Where the currents descended, the lithosphere would be dragged downward into the asthenosphere and many earthquakes would occur. If the oceanic ridges represented areas of spreading, the anomalies of temperature and sediment found in their vicinity would be explained. The old sediment on either side of the ridges had spread, atop the moving oceanic crust, away from the area where it had initially rested.

Evidence to prove this sea-floor spreading was ingeniously deduced by the English geophysicist Frederick Vine. He used data showing that throughout geologic history the earth's magnetic field has switched many times. That is, what we now identify as the north magnetic pole was once actually the South Pole, and our compasses would have pointed south in locating the pole. Sediments containing iron that had been deposited during a time when the poles were reversed would reflect the change: instead of being aligned to our current north-south magnetic field, they would be reversed. Vine thus reasoned that rocks formed on either side of the ocean ridges show bands of magnetic fields and that rocks representing a magnetization direction on one side of the ridge should be seen on the other as well. His theory proved correct.

Putting together all of this information, we have a picture of great crustal planes, each moving as a distinctive unit atop the asthenosphere. Figure 13-3 diagrammatically represents these events. The crustal plates are spreading apart and giving rise to volcanic activity in, for example, the oceanic ridges. Evidence for the spreading is seen in the magnetic stripes on either side of the ridges. Where two crustal plates coverage, compression occurs, and one plate plunges below another. Such an event is represented by the great oceanic trenches that border some continental areas. In areas of spreading and compression the crustal instability gives rise to earthquakes and volcanoes.

Figure 13-3. Schematic diagram of the basic motions associated with plate tectonics.

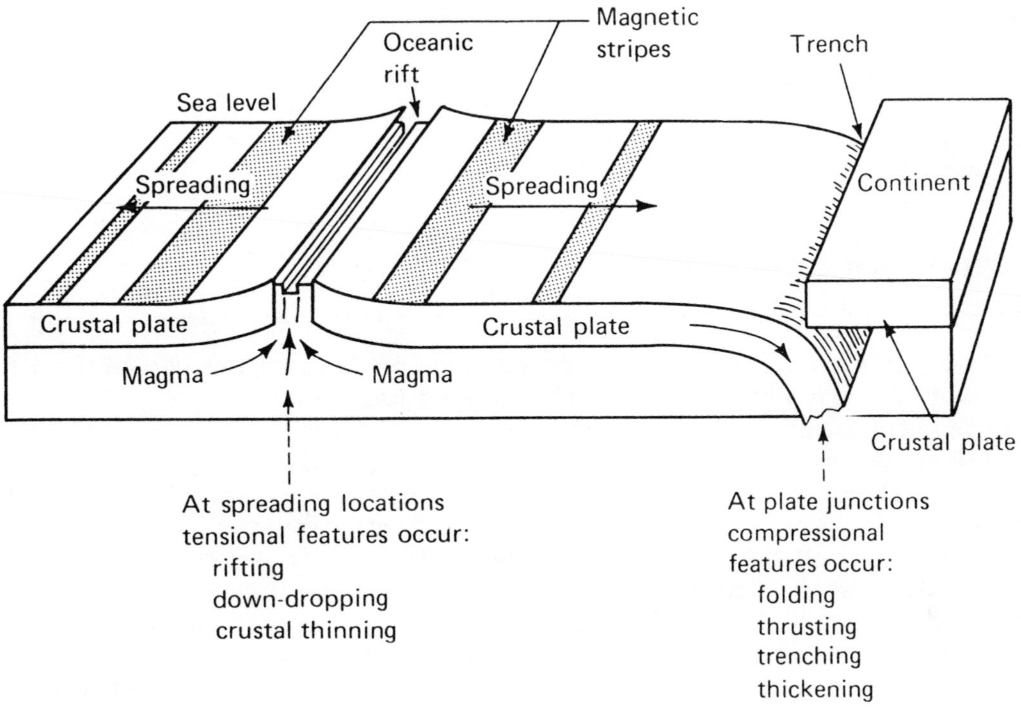

Identification of these plates on a map of the earth is facilitated by locating the areas of crustal instability. In outlining the limits of individual plates, a number of scales can be used, and, as a result, anywhere from six to twenty plates have been identified. The boundaries of the plates mark areas where spreading occurs (for example, along the East Pacific or Mid Atlantic Ridges) or where one plate plunges beneath another (for example, along the west coast of the American or the east coast of the Eurasian plate). The unified theory behind the new look at crustal motion is aptly termed plate tectonics, plate referring to the individual earth masses identified, and tectonics, the movement and deformation of the crust on a large scale. Evidence of the importance to people of plate tectonics is seen in many areas of the world, principally in volcanic eruptions and earthquakes. Earthquakes result from the thrusting of one plate beneath another or the sliding of one plate against another. In fact, all major forms of earth motion, and the frequency with which that motion occurs, can be related to the movement of the giant plates.

PLATE TECTONICS AT WORK

The effects of plate tectonics can be seen in many parts of the world: in Washington, with the eruption of Mt. St. Helens; in California, with the active San Andreas fault; and in Surtsey, a volcanic island south of Iceland. The number of earthquakes in Turkey and Iran, for example, is associated with movement between the Eurasian and African plates. The massive earthquake that devastated part of Mexico City in 1985 was also associated with movement along plate boundaries. The Mexico City earthquake measured 8.1 on the Richter scale, a measure of ground motion as recorded on a seismograph. Every increase of one number in magnitude (e.g. 4 to 5) means the ground movement is approximately 10 times greater (although the energy released may be 30 times greater).

The earthquake most destructive of human life was in Shansi Province, China in 1556 (magnitude unknown) which killed 830,000 people, [1], although China came close to that again when

650,000 perished in a magnitude 7.6 earthquake in Tangshan Province in 1976.

Despite the significant correlation between earthquakes and plate boundaries, the three most widely felt earthquakes in North America occurred in the central portion of the North American plate. These earthquakes, with estimated magnitudes of 8.6, 8.4, and 8.7 occurred in the mid-Mississippi River Valley near New Madrid, Missouri. These deep-seated earthquakes, believed to be fractures in the North American plate, caused plaster to crack in buildings as far away as Richmond, Virginia. Fortunately, the central region of the United States was not densely settled at that point in history.

In explaining why such movements occur, advocates of plate tectonics were faced with the same problem as Wegener: What causes the motion? Again, although aided by many more observations and more accurate facts about the earth, scientists still disagree on the exact cause. A number of conflicting models have been produced, but all used, as the causative mechanism, the idea of convective currents in the mantle. The models vary in their location of the currents. Some assume they are restricted to the upper limits, while others say they are in the deep mantle. One possible source of energy to create the convective currents is heat released by radioactive decay. The decay of uranium, potassium, and thorium could, in sufficient time, produce enough heat energy to create such currents. Another theory suggests that the plates, which are cold surfaces, descend into the warmer trenches in the mantle and pull the remaining portion of the plate after them. It is not yet possible to identify with accuracy the basic cause of plate movement, but with data derived from ongoing research projects the answer may eventually become clear.

Movement and fracturing of continental plates also allows molten rock (magma) from the earth's interior to rise towards the earth's surface. Most of this magma is extruded through linear fissures or fractures. Approximately 81 percent reaches the earth's surface in this manner. [1] The other 19 percent of the magma rises at particular points rather than along linear fissures, some of it contributing to formation of the more familiar central-vent volcano.

Volcanoes at the plate margins have lava rich in silica which contributes to its high viscosity (it does not flow easily) and build up of explosive cases is to be expected. Consequently, eruptions from these volcanoes are extremely violent. Ash and dust form cauliflower shaped clouds above the volcano, and hot ash can create extreme devastation to the land surface. In 1980, Mount St. Helens in Washington state provided an excellent example of the power of volcanism within the region of plate boundaries. Within a few minutes after Mt. St. Helens' eruption, the ash plume had risen some 20 kilometers (12 miles) into the atmosphere.

CONCLUSION

It is not unreasonable now to assume that, as parts of more extensive plates, continents do drift. The modern interpretation differs from the one proposed by Wegener, but it is clear that Wegener was basically correct in his viewpoint. It is important to note, however, the differences in the ways in which his continental drift theory and the modern plate tectonics theory have been derived. Wegener worked alone: he observed, collected data, and drew conclusions. On the other hand, the idea and development of plate tectonics theory does not belong to a single individual but grew out of research teams and from sequential research studies. Such a difference reflects the evolution of research in the natural sciences. Of course, brilliant scientists still develop theories single-handedly, but the earth sciences, like all branches of science today, have become so complex that researchers have had to specialize in limited areas. Thus, the theory of plate tectonics grew out of a collaboration of specialists in earth magnetism, deep sea drilling, and earthquakes.

SUGGESTIONS FOR FURTHER WORK

1. Compile a list of major volcanic eruptions of the past fifty years and plot locations on a world map. How do the locations relate to the distribution of global plates?

2. There is a common notion that part of Southern California will break away from the mainland and drift off to sea. In the light of the theory of plate tectonics and the location of the San Andreas fault, explain how such an idea came about. Do you think the notion is correct?

SELECTED SOURCES FOR FURTHER READING

Condie, K.C. Plate Tectonics and Crustal Evolution, New York: Pergamon Press, 1982.

Paton, T.R. Perspectives on a Dynamic Earth, London: Allen & Unwin, 1986.

REFERENCES

1. Smith, P., ed. The Earth, New York: Macmillan Publishing Co., 1986.
2. Wegener, A., The Origin of Continents and Oceans, translated from 3rd German ed. by J.G.A. Skerl, New York: Dutton, 1924.

Chapter 14

Energy from the Earth

The Context

Energy generated within the earth usually manifests itself in violent ways. Earthquakes and volcanoes provide spectacular displays of this energy, often with dire consequences to human beings. But just as people have harnessed the energy of rivers and tides, of sunlight and the atom, so they are attempting to control energy from the earth.

The use of this earth heat, or geothermal energy, presents unique problems. Since the conditions that create geothermal sources are associated with crustal instability, the distribution of the resource is limited to geologically active portions of the earth's crust. A second problem concerns the engineering aspects of harnessing geothermal energy. In this chapter we shall examine some of the areas where geothermal energy is actually used. Volcanoes and earthquakes are an integral part of the environment in these cases, and as yet we have little technology to control and direct such a powerful and potentially destructive resource.

The geyser, Old Faithful, in Yellowstone National Park is one of the most spectacular natural phenomena in the United States. Until quite recently its eruption of hot water and stream jetting to a height of some 30 meters (100 feet) could be predicted to within a few seconds. In recent years, however, Old Faithful has become somewhat less faithful, for a nearby earthquake and erosion of the geyser tube have affected the geyser's activity. It remains, nonetheless, a spectacular display of earth's forces.

Old Faithful actually represents a "leak" in the heating system of the earth. Such leaks are relatively rare. There are only three main geyser areas in the world: Yellowstone National Park, Iceland, and the Taupo-Rotorua district of New Zealand. They are all a result of an unusual combination of factors that are essential for their existence. In the first place, there must be a source of subsurface heat, such as the deep rock heated by magma, the molten rock that occurs far below the earth's surface. Then, an extensive system of rock fractures, or joints, must extend from the warm area of subsurface rock to the surface. A local high water table is also necessary to fill the entire joint system, from the heated area to the surface, with water. The high water table causes the water in the lower parts of the joints to be under considerable pressure because of the weight of water above. Water under pressure can be heated to temperatures in excess of 100°C (212°F) without boiling. However, heat does pass up the joint system, and the resulting expansion causes the water to overflow at the surface and reduces the pressure on the hot deep-seated water. With pressure relieved, the overheated lower water changes rapidly to vapor. The vaporization increases the volume some 1,700 times as the geyser erupts. Vapor, steam, and water are violently ejected until the pressure is equalized. As the eruption subsides, the system is slowly replenished with water and the cycle is repeated.

Although there are few places in the world with all the conditions necessary to create geysers, there are many locations where heat sources exist. In some cases the subsurface heat creates hot springs. Such water has been used since Roman times for hot baths and health treatments. The earth's heat, or geothermal energy, responsible for these springs of hot water appears to provide an additional source of energy to help meet current demands.

ICELAND'S THERMAL ENERGY

There are few places on earth that can match the physical geographic variations found in Iceland (Figure 14-1). In this island of 104,000 square kilometers (40,000 square miles) there exist, side by side, extensive glaciers on land that, by its absence of vegetation and texture, is almost moon-like in appearance, and areas where steaming springs provide a warmth that seems totally out of place in this northern realm. The amazing variety is a result of Iceland's location: its latitude and its position in relation to the Mid-Atlantic Ridge. The latitude, with northern fringes of the island bordering on the Arctic Circle, is responsible for an environment in which glaciers still occupy 12 percent of land area. These glaciers are probably the remains of much more extensive ice sheets that once covered the island in Pleistocene times. Iceland is the largest exposed piece of land on the Mid-Atlantic Ridge, which represents a spreading portion of the earth's crust. Like other exposed parts of the ridge--the Azores, Saint Paul Rocks, Ascension Island, St. Helena, and Tristan da Cunha--Iceland experiences frequent geological activity. Small shocks are frequently recorded in Reykjavik, the capital. Volcanoes, like the celebrated Mount Hekla--the supposed abode of the damned in the Middle Ages, and more recently-formed Surtsey--are an integral part of the environment. Set amid this activity are more than 800 hot springs and a number of geysers. Such a mixture of hot and cold can lead to quite a special phenomenon. One problem is great floods of water known locally as jokulhaups. These result from volcanic activity beneath ice sheets. Huge volumes of ice are melted beneath the ice surface, but the water is contained until a rupture occurs along the edge of the glacial ice. Billions of cubic feet of water are then suddenly released, "carrying huge ice fragments and boulders at velocities of about 60 miles [100 kilometers] per hour. Farmhouses, barns, people, and livestock in the flood's path are all swept up in a maelstrom from which virtually nothing escapes." [2]

Figure 14-1. Map of Iceland showing the coexistence of glaciers and hot springs.

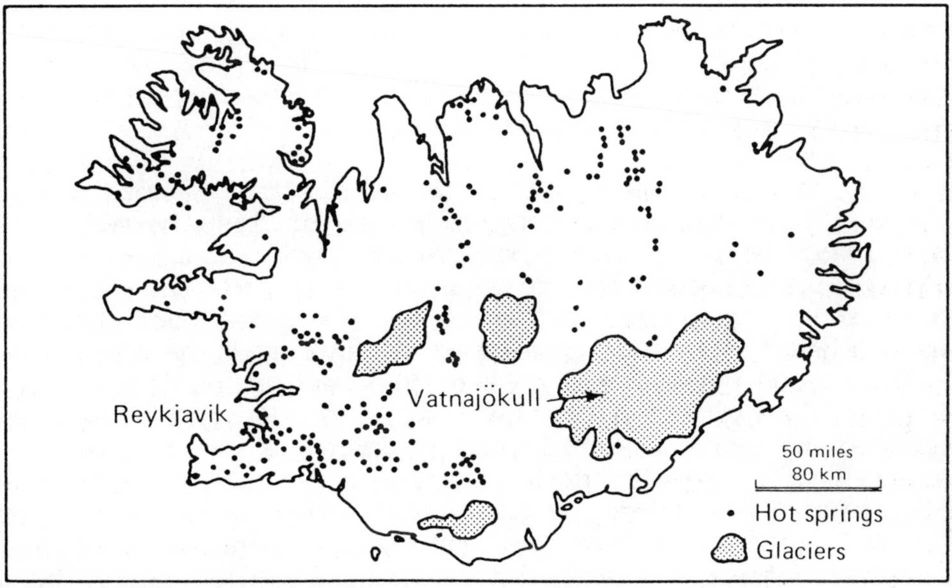

Volcanic activity is another problem in Iceland. Heimaey, located in the Vestmann Islands off the south coast of Iceland, erupted without warning on January 23, 1973 (Figure 14-2). Within six hours of its eruption almost all of the 5,800 residents of the island had been evacuated to the mainland. By the end of February a cinder cone had grown to more than 61 meters (200 feet) above sea level, while lava flowed in a general northeasterly direction toward the port of Vestmannaeyjar, Iceland's major fishing port. Over the following months, parts of Vestmannaeyjar were destroyed, and lava began filling the harbor. In order to stop the lava, water was poured on the flow, which cooled the temperature of its surface and retarded its movement. This method had been used experimentally, but the Heimaey eruption was the first time it was used on a large lava flow. The results, as reported by Williams and Moore, were basically successful from both the economic and scientific viewpoints [5]. The cold seawater was used to wet the lava, and the sea gave easy access for the necessary pumping equipment. The lava itself was slow moving, and the flow was controlled. About half of the original population returned to Heimaey, and the fishing fleet out of Vestmannaeyjar continues its activities.

There are also benefits from the strange hot and cold mixture, for Iceland has long made use of its thermal energy. Since 1925, it has drawn upon natural hot water for domestic heating, for greenhouse cultivation, and for the heating of public swimming pools. The use of natural hot water for domestic heating is best developed in Reykjavik, where the homes of some 85,000 persons and most of the central business district use it [3]. To supply the heat, the Reykjavik system is connected to more than 100 wells, many of which are in the thermal area of Reykir, some 16 kilometers (10 miles) northeast of the city. Water leaving the wells ranges in temperature from 80°-140°C (about 176°-284°F) and, after passing through the main central heating plant, enters homes at temperatures between 75°-80°C (approximately 167°-176°F). The water is eventually discharged from the system when its temperature falls to 35°C (about 88°F). The use of this heating source saves some 70,000 metric tons (77,000 U.S. tons) of fuel oil every year, a factor

Figure 14-2. Extent of lava flows on Heimaey in the Vestmann Islands. Inset shows location in relation to Iceland.

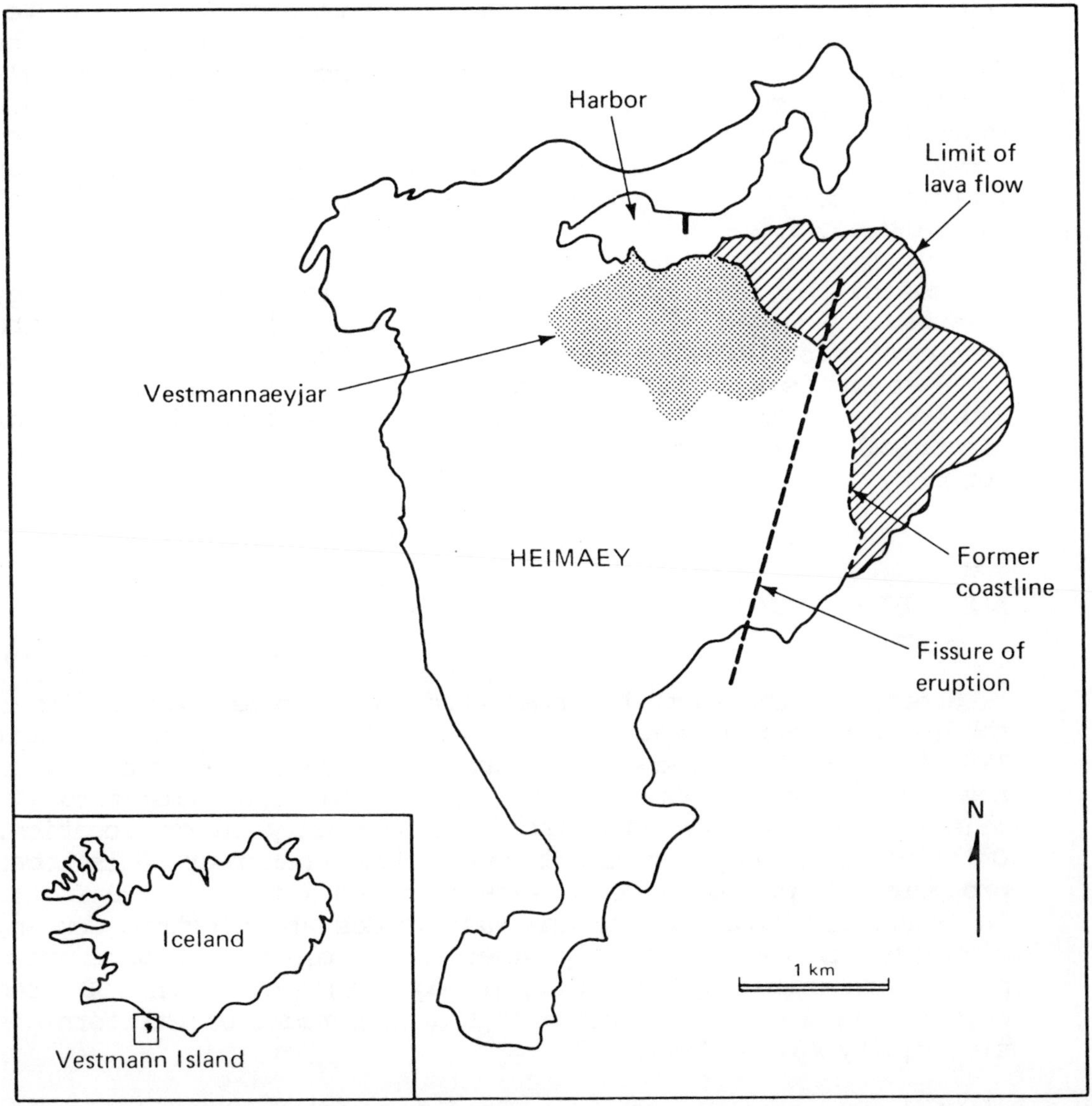

of appreciable significance in these days of rising fuel costs.

Geothermal energy is also used to heat greenhouses. Although Iceland is surrounded by the sea, which moderates its climate, the growing season is very short. Thus, greenhouse production plays a significant role in supplying some of the country's vegetable needs. The small town of Hveragerdi, located some 330 kilometers (200 miles) from the Arctic Circle, has achieved local fame for its production of fruits, vegetables, and even tropical flowers. Geothermal energy is also used to heat swimming pools.

Iceland has clearly made good use of its geothermal energy sources. However, it has not, as yet, fully developed the resources for generation of electricity. This lack of development is, at least in part, due to the availability of hydropower on the island. But the use of geothermal energy for electrical production is taking place in other parts of the world.

ELECTRICITY FROM THE GEYSERS

Steam from the earth has been used to generate electricity in the Larderello region of Italy since 1904, but it was not until 1958 that geothermal energy was used to generate electricity in the United States. The Geysers area, about 120 kilometers (75 miles) north of San Francisco, is the only major location. Drilling in the Geysers area began in 1955 when four wells began producing steam from a depth less than 300 meters (1,000 feet). In 1968 a drilling company and Pacific Gas and Electric Company signed a contract to supply steam to an electrical generating plant, and commercial production began in 1970. In 1987 the Geysers field was supplying more than 2 percent of California's electricity [3:361-364].

Steam from the wells is collected, filtered, and passed through turbines. Engineering adjustments were needed at the generating plant because of the relatively low pressure and temperature of the steam. In a typical power plant using fossil fuels, steam has a pressure in the order of 210 kilograms per square centimeter (3,000 pounds per square inch (psi)). At the Geysers it is only about 7 kilograms per square centimeter (100 psi). Similarly, the Geysers steam is only 205°C (400°F) compared to the 550°C (1022°F) at typical power plants. Because of low pressure and temperature of the natural steam, the Geysers plant is less efficient than conventional ones. Operating at about one-third the efficiency, the Geysers plant requires 450 megawatts of steam to produce 100 megawatts of electricity. Such a disadvantage is somewhat offset by the relatively low construction and operating costs of the plant. Its success can be measured by the fact that the utility company intends to add new generating capacities each year for several years. The initial capacity was 12,500 kilowatts, and it has been increased many times. How long production can continue is a question, for it is not known how long the steam from a given well will last, although a lifetime of 100 to 200 years is projected.

Underground wet steam deposits are more common but harder and more expensive to convert to electricity. These deposits contain water under such high pressure that its temperature is far above the boiling point of water at normal atmospheric pressure. Four small scale plants utilizing this type of geothermal energy have been built in the United States, but suffer from metal corrosion and drilling problems [3].

POLLUTION HAZARDS

It might be thought that geothermal energy is a clean energy. Its production does not involve the byproducts associated with burning coal or oil, nor is there any radioactive waste as in the case of nuclear power production. In fact, geothermal energy production does have its own pollution problems, which depend on the amount of water in the geothermal steam and the chemical content of that water. The Geysers system is low in pollution, for the steam utilized is not accompanied by liquid water. Such "dry" steam is rare; most geothermal wells produce a mixture of water and steam. These components need to be separated before electricity can be generated, and the problem of water disposal arises. This disposal is often made more difficult because the water usually contains salts. To simply discharge this waste into surface streams would play havoc with the ecologic balance.

Because of the problem of surface water pollution, most design projects for geothermally generated electricity assume that waste water will be disposed of by reinjecting it into deep wells. This is the process followed at the Geysers. Exhaust steam from the turbines is condensed and trapped in cooling towers where a good deal of the water ultimately evaporates. The rest of the water that accumulates contains ammonia and boron. Discharging such water to local streams would result in severe pollution problems; so the water is reinjected into deep wells.

However, mineralized water is not the only pollution hazard. As noted in the preceding paragraph, steam and heat are released to the atmosphere. Given the large amount of steam required to produce a significant electrical output, the amount released can be appreciable. Although the atmosphere is generally considered more capable of handling such pollution than lakes and rivers, the effects of heat release on local weather have not been clearly defined.

Noxious gases are also emitted by geothermal plants. One study indicates that the quantity of sulfur released to the atmosphere at the Geysers plant is the equivalent of that emitted by a similar size plant using low sulfur fossil fuels [1]. It is known, too, that the evil-smelling hydrogen sulfide does escape to the atmosphere. The construction of geothermal plants, like any other power plants, needs a careful control of emission standards.

THE FUTURE AND CONCLUSIONS

Development of geothermal energy in the United States will largely occur in the geologically active portions of the west and in Hawaii. The United States Geological Survey estimates that as much as 525,000 hectares (1.3 million acres) of land may eventually provide an energy source of from 15 to 30 thousand megawatts.

There is clearly a future for geothermal energy, and researchers are investigating all possible approaches to its use. For example, they have been attempting to use geothermal heat that is not in contact with underground water systems: the so-called hot rock. One plan suggests that artificial cavities, made by conventional or nuclear explosives be created in the hot-rock area. Water from the surface could then be circulated through this heat source. But despite the great promise of geothermal energy, its use as a resource needs to be considered in relation to total energy needs. We have seen that geothermal energy can only supply a small part of the projected energy needs of the United States. Nonetheless, it is a source that will be developed, and, as part of that development, we shall learn a great deal about the geochemistry of the earth.

SUGGESTIONS FOR FURTHER WORK

1. Investigate the advantages and disadvantages of the use of geothermal power in the Salton-Mexicali area on the American-Mexican border.
2. The chapter mentions a number of other islands located on the Mid-Atlantic Ridge. Select any one of these and compare its physical environment to that of Iceland.

SELECTED SOURCES FOR FURTHER READING

Armstead, H. Geothermal Energy: Its Past, Present, and Future Contributions to the Energy of Man, London: E. and F.N. Spon, 1983.

Miller, G.T. Living in the Environment, Belmont, California: Wadsworth Publishing Co., 1988.

REFERENCES

1. Goldsmith, M. Geothermal Resources in California: Potentials and Problems, Pasadena, California: California Instutite of Technology, Environmental Quality Laboratory, Report No. 5, 1971.
2. Kane, J. "Iceland's Thermal Geology," Natural History, Vol. 78, No. 1 (January, 1969).
3. Miller, G. Living in the Environment, Belmont California: Wadsworth Publishing Co., 1988.
4. U.S. Geological Survey. Briefs on Geothermal Energy, reprinted in G.O. McKenzie and R.O. Utgard. Man and His Physical Environment. Minneapolis: Burgess, 1972.
5. Williams, Jr., R.S. and J.G. Moore. "Iceland Chills a Lava Flow", Geotimes, Vol. 18 (1973).

Chapter 15

Problems With Slopes

The Context

Physical geographers have always been interested in differnt kinds of landforms. Part of that interest has centered on the surfaces that separate higher from lower land: the sloping sides of hills and mountains. The study has not proceeded without differnences of opinion, and even today there is controversy over the formation and retreat of slopes. In recent years the study of slopes has become environmentally significant. The use of hillsides for recreation, the development of buildings on steeply sloping land, and the potential for landslides that arises from such uses indicate the knowledge of the basic nature and properties of sloping land is essential.

The cases we still examine in this chapter concern the use, or misuse, of sloping land and the disasters that can result. First we will look at how and why natural avalanches and landslides happen. Then we will consider the ways that peoples' activities affect slopes and create hazardous conditions.

NATURAL AVALANCHES AND LANDSLIDES

Each winter approximately 10,000 avalanches are observed in United States mountains. Compilation and analysis of data associated with each avalanche provide a way to help mitigate future disasters. One technique is to use explosives to trigger an avalanche deliberately while the slope is clear of people and thereby to remove an unstable accumulation of snow and reduce the hazard (Figure 15-1).

The United States Department of Agriculture Forest Service has published a highly instructive and fascinating account of avalanches in the United States. The following summary along with Figure 15-2 illustrates how a snow slide happens.

On January 29, 1970 a large snowstorm dropped 950 millimeters (about 37 inches) of snow in the Wasatch Range in Utah, increasing the snow depth from approximately 190-260 millimeters (74-101 inches). Patrols performed avalanche tests on the ski slopes in the Peruvian Ridge area by throwing hand charges and making ski cuts. There seemed to be no danger of an avalanche, so a small party that was making a promotional film prepared to ski down the Lower Peruvian Ridge. As the two ski instructors, Lundberg and Engen, descended toward the two photographers further down the slope, the snow fractured above and around them.

Figure 15-1. As part of the U.S. Forest Service's avalanche control program, this avalanche was released by a 75mm shell exploded in a suspected wind slab area. Wasatch National Forest, Utah. (Photo courtesy of National Agricultural Library, Forest Service Photo Collection).

Lundberg didn't see the snow fracture but knew. . . he was having trouble keeping his balance. He then realized he had been caught in an avalanche and struggled to keep his feet. He maintained his balance, skied with the slide, and came to a stop at the edge of the moving snow.

Engen saw the snow fracture out of the corner of his eye. A powerful skier, he pointed his skis straight downhill trying to ride the slide out. With a lot of

Figure 15-2. Diagram of events associated with the snow avalanche described in text (From U.S.D.A. Forest Service, General Technical Report 8, 1975, p.109.)

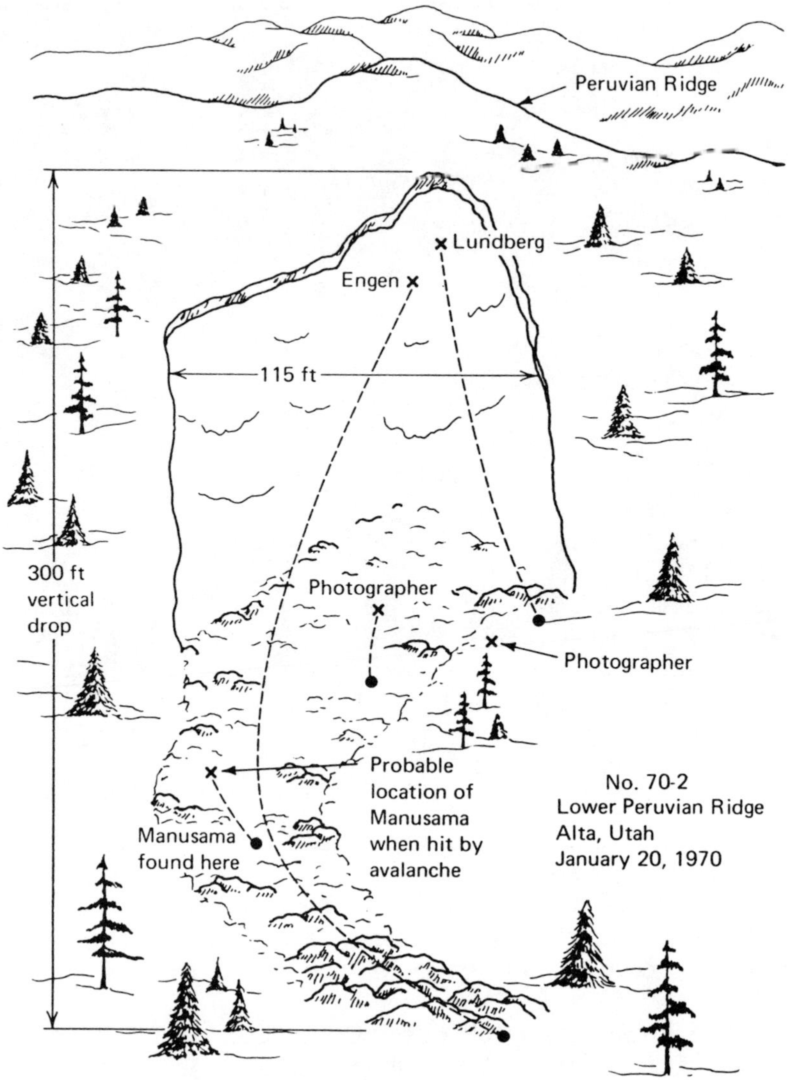

snow above him overtaking him, he had trouble standing up in the churning mass. He gained speed quickly and zoomed downhill and out of sight as the avalanche turned a corner below a small knoll [4:108-110].

One of the photographers was directly hit by the snow and was buried up to his neck. The other photographer was out of the way. There was one death in that avalanche. Unnoticed by the film party, a bystander had been caught in the snow. He was not missed and found until almost six hours later. "He was located under 3 feet [one meter] of snow, head downhill and no indication of an ice mask around the face. All his ski equipment was intact, indicating he had been carried only a short distance by the avalanche" [4:110]. He he died from asphyxiation.

The avalanche described here represents a rather special form of problem associated with sloping land. It comprises the downhill movement of a mass that is not actually a part of surficial earth material and rock. But, like snow, surficial material does move, although at variable rates. In some cases the downhill movement is similar to the snow avalanche, and great masses of material suddenly slide downward at high speeds in excess of 160 kilometers (100 miles) per hour. In other cases the movement is slow, sometimes barely perceptible to the observer from day to day. Slopes most susceptible to sudden downhill movement typically exhibit certain characteristics. Thus, they usually are more than 150 meters (about 500 feet) in height, with a slope angle of more than 25 degrees. In addition they are likely to be composed of fractured rock which has been undercut by fluvial or glacial erosion and to have at least one other characteristic of low strength or potential instability [1: 1288-90]. Potentially unstable slopes, however, may resist the pull of gravity indefinitely, so the triggering mechanism constitutes the next aspect to consider.

Theoretically, the motion is easily explained, for downhill movement occurs when the cohesion of the surface material is less than the gravitational force. If a rock is not attached to the underlying stratum, there is no cohesive force, and such a rock would easily slide downhill. If the rock is joined to the under-lying stratum, its molecular cohesion to that stratum would make it less likely to move. In order for the attached rock to move, some force has to rupture the molecular cohesion.

On Good Friday, 1964 south-central Alaska experienced a massive earthquake. So great was the movement that an area to the west of the active zone, comprising some 78,000 square kilometers (30,000 square miles), was lowered as much as two meters (about 6 feet). To the east an area of 130,000 square kilometers (50,000 square miles) was elevated above its former height, in some locations by as much as 10 meters (33 feet) (Figure 15-3). The shock of the earthquake jarred large masses of earth into motion, and the resulting landslides and slumps were responsible for destroying buildings, communications, and utility lines. At Anchorage four large slides caused extensive damage to buildings. They resulted from failure of cohesion in water-laden bluffs made of soft, silty clays. The earthquake shocks also caused submarine landslides. At the seacoast towns of Valdez and Seward near Anchorage the violent shaking associated with the earthquake caused settling of the delta materials on which parts of the towns were built. Slumping of the submarine deposits, which began well below sea level, carried away the waterfront areas of both towns.

The destruction of the Peruvian villages of Calla and Ranrahirca resulted from the downhill movement of more than 2.7 million metric tons (3 million U.S. tons) of debris. In this case the movement was not triggered by an earthquake, but rather by the fracturing of a glacier on the flanks of the 6,800 meter (22,205 foot)-high mountain, Nevada Huascaran. In January, 1962 millions of tons of ice broke away from the leading edge of the glacier and began a seven-minute, 14.5-kilometer (9-mile) long run down to the valley floor. The moving ice scoured the land surface and became a hundred kilometer- (60-mile)-an-hour torrent of ice and water mixed with debris ranging from huge boulders to disintegrated buildings and trees torn up by their roots. The torrent's zigzag motion down the mountain side caused complete destruction of some villages, but by bouncing off the uneven sides of the valley walls, it left others quite untouched. This disaster resulted in the death of 3,500 Peruvians.

Figure 15-3. Elevation changes resulting from the 1964 Alaska earthquake.

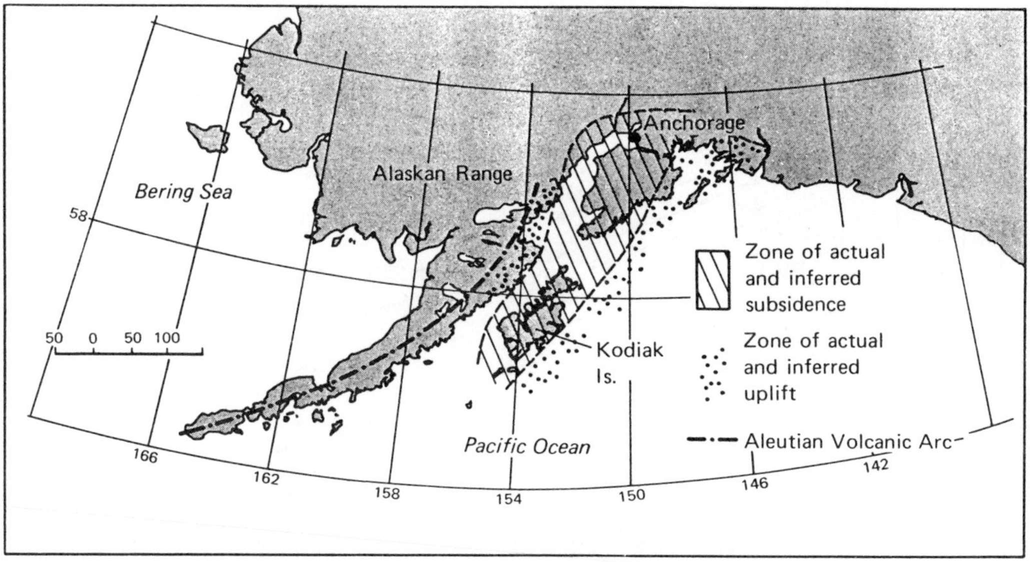

A third example of slope failure leading to rapid downhill motion comes from the intensively studied landslide that occurred in Gros Ventre, Wyoming in June, 1925. This older example is used because it represents the type of geologic structure that frequently gives rise to landslides. The slide took place in the valley of the Gros Ventre River some 56 kilometers (35 miles) south of Yellowstone National Park where the rock formations, shown in Figure 15-4, slope in a northeasterly direction at angles between 18 degrees and 21 degrees. In the spring and early summer of 1925 heavy rains and melting snow in the Gros Ventre Mountains caused clay layers just above the carboniferous stratum to become totally saturated. The moisture-laden clay lacked cohesion with the rocks below and, as a result, the rocks above the clay-limestone junction suddenly slid downhill. The great mass, estimated at 39 million cubic meters (50 million cubic yards), carried with it a dense pine forest, and when the debris eventually came to rest it completely blocked the Gros Ventre River. The river

Figure 15-4. The Gros Ventre landslide illustrates how movement can occur along strata that become saturated with water. (Based upon original research by W.C. Alden.)

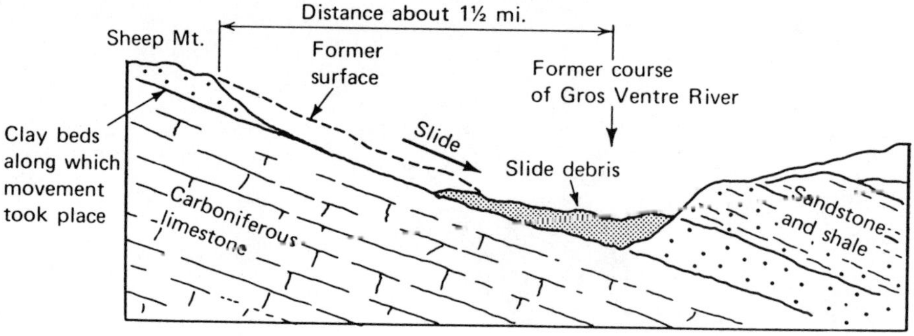

was in flood at the time, and a large lake formed. So rapidly did the water rise behind the dam of slide debris that in less than 24 hours a ranch house, located about 18 meters (60 feet) above the former level of the river, floated off its foundations.

These examples show rapid movement on slopes. Slower rates of movement also cause landform changes. Although cohesion retards downhill movement of surficial material, the effect of gravitational pull is always at work and, over time, may lead to a very slow creep of material downhill. The effect of this creep is seen in many ways. Figure 15-5 shows some typical examples: retaining walls bulge, telephone poles and fence posts tilt, and gravestones topple over. The diagram also shows how such movement is related to the bending of rock layers near the surface. While cohesion stops the surface material from sliding downhill, the gravitational pull, over a long period of time, slowly deforms the rocks.

Figure 15–5. The downhill creep of soil and weathered rock is seen in the movement of many structures.

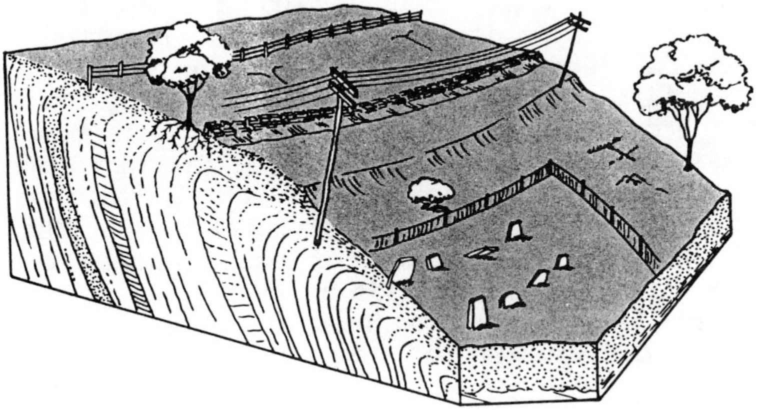

THE USE OF SLOPING LAND

The foregoing examples demonstrate that, under natural conditions, steep slopes are often unstable. In many cases, however, the stability of a slope is maintained by existing environmental conditions, such as vegetation cover or movement of groundwater away from lines of potential movement. When a new, disturbing force, such as farming or building, is added, the existing equilibrium is modified and the potential for downhill movement may be considerably enhanced. Agricultural use of sloping land has shown that the clearing of land on slopes results in erosion and the loss of vast quantities of topsoil. Drawing upon past experiences, most farmers now follow simple landuse guidelines. Strip planting, contour plowing, and slope breaks are integral parts of farming on slopes. Unfortunately, the realization that the use of steep slopes can lead to problems seems less apparent to land developers than to farmers. Buildings continue to be constructed on land that is really too unstable to support them (Figure 15-6). Examples of devastation that can result from such misuse are plentiful in California. The prestige of owning a home that offers a panoramic view from its windows has resulted in house construction on steep slopes high above the base of the slope. In constructing such a building, slope vegetation must be cleared and access roads provided. Houses must be anchored to the hillside literally to stop them from falling downhill. All of these modifications result in changes in surface cohesiveness and in alterations in the system of water flow. Thus, after a torrential rain the foundations of such houses are destroyed and the houses themselves slide down the slope. While the excessive rain is often blamed for the slide, it is evident in the long run that misuse of land is the real cause of the disaster.

Figure 15-7 shows two examples of the types of problems involved. The upper diagram shows common causes of hill slides where homes were built on unstable land. The original surface was altered by cut and fill, with the home located on flat, filled land and a steep slope leading away downhill. The foot, or "toe," of the slope was undercut, and this, together with the loading of material at the top of the slope, resulted in a landslide. The lower diagram, taken from a study by Leighton, shows an actual case where even a well-engineered house was lost [3:97-132]. The reason for its destruction was slippage along the clay seam. Disturbance during construction of the house allowed water to find its way to the clay seam, which it lubricated. When excavations for another home began farther down the slope, loss of support for higher material resulted in a slide along the clay seam.

Such events occur in many parts of the United States. In Colorado, for example, development of sites high on the slopes of the Rockies, where unpolluted air, privacy, and scenic vistas make the land very attractive, has led to enormous problems. Most of the water runs off; so there is competition for groundwater. Since there is little soil on the slopes, septic tanks do not work well and downhill seepage contaminates wells,

Figure 15-6. Construction of homes on unstable slopes can have dire consequences. (Photo by arrangement with U.S. Department of Agriculture, Soil Conservation Service).

ponds, and streams. Foresters too find fault with such locations, for the limited and often tortuous access roads make hillside homes difficult to reach should fire break out. Fire, of course, moves faster uphill than on flat land, and the fire hazard is a constant concern.

Figure 15-7. Examples of the way in which use of slopes can create problems for houseowners. (From R. Beach Leighton, "Landslides," in Geologic Hazards and Public Problems, Conference Proceedings, Office of Emergency Preparedness, Region 7, Santa Rosa, California,1969.)

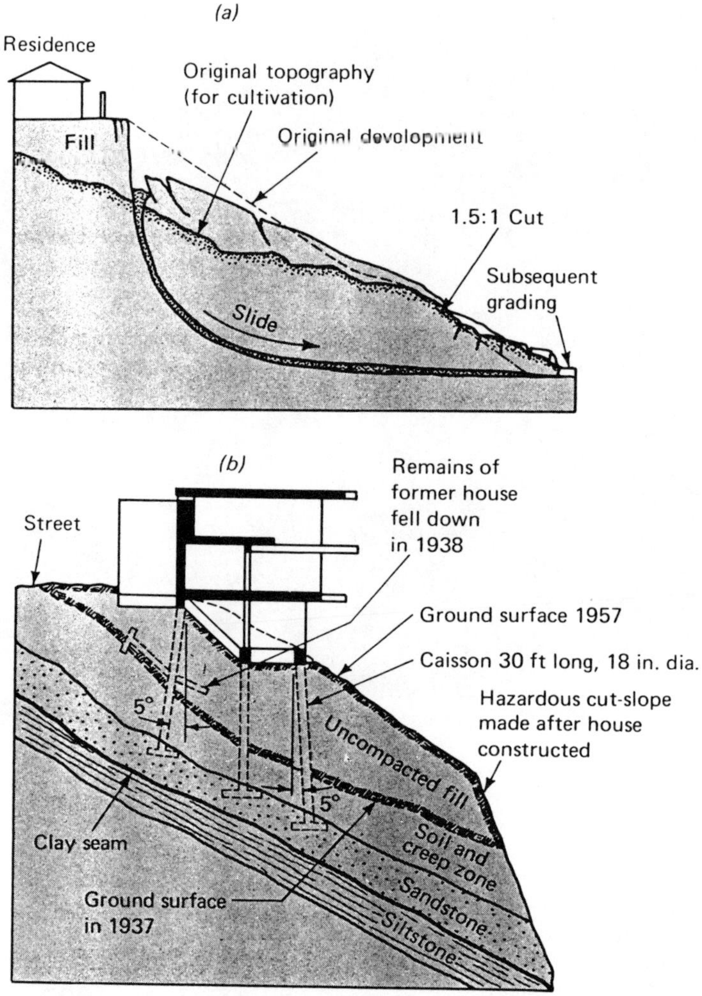

Despite the plentiful supply of problems with slopes in the United States, I have chosen to look at a case from Italy in detail. It provides one of the few instances in which site developers have been criminally charged for a so-called natural disaster.

THE VAIONT EVENT

As we have seen, a landslide can result in terrible destruction and loss of life. If the landslide debris plunges into a reservoir, its effects go far beyond the immediate vicinity of the landslide, for the displaced water rushes downstream. This is what happened in Italy in what is known as the Vaiont Disaster.

The Vaiont River, a tributary of the Piave, has an interesting geologic history. Sedimentary rocks formed during Jurassic and Cretaceous times were folded to create a syncline in which the ancient Vaiont flowed. With the onset of the Pleistocene, the valley was occupied by a glacier, which scoured it. Ice retreated some 18,000 years ago, and again the valley became the course of the Vaiont River. Since the river was reestablished it has carved a deep valley some 200-300 meters (650-1,000 feet) below the level of the glacial valley (Figure 15-8).

Rocks comprising the valley responded in a number of ways to the changing events. Most importantly, fractures formed in the rock through two processes. The first process created rebound joints or fractures that resulted from a readjustment to pressure differences caused by the removal of a heavy mass of ice in which the rocks rebounded after having been compressed by the weight of ice. A second set of joints was formed by the rapid downcutting of the postglacial Vaiont. This process again represents a readjustment of the rock to changing pressures. In this case the rocks originally supporting the sides of the valley had been eroded, and the rock that remained had cracked along lines of weakness.

Figure 15-8. block diagram illustrating the old glacial and the newly carved river valleys of the Vaiont. Two sets of joints contribute toward instability of the valley walls. (Adapted from G.A. Kiersch, "Vaiont Reservoir Disaster," Geotimes (May-June, 1965),p.11.)

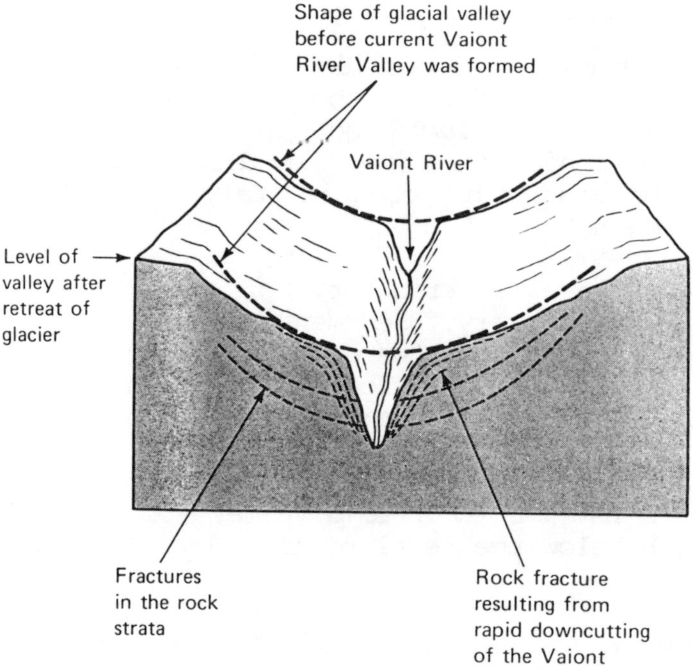

The weakness of the fractured walls of the canyon was made even worse by the fact that the rocks themselves were inherently weak, and the limestone strata of the group had been weakened through solution. These factors, plus the steep walls of the valley, created a situation that was right for landslides. Nonetheless, a dam was constructed across the valley and a reservoir formed. There was a warning of trouble in 1960 when a slide involving some 700,000 cubic meters (900 cubic yards) of debris occurred on the left bank of the reservoir near the dam. Following that slide, limits were placed on the amount of water that could be retained in the reservoir. However, in September of 1963 observation points around the reservoir showed signs of creep, but not until early October did engineers realize that an enormous mass of material was unstable. Their immediate response was to lower the level of water in the reservoir some more, but the heavy rains of the season kept the water level high. Then, the slide occurred.

> The quick sliding of the tremendous rock mass created an updraft of air with rocks and water that climbed more than 260 meters [850 feet] up the right canyon wall from reservoir level. Subsequent waves of water swept over both abutments, some 100 meters [330 feet] above the dam crest. The surge of floodwater down Vaiont Canyon swept away two high level bridges and was over 70 meters [230 feet] high at the confluence with the Piave Valley, about a mile [1.6 kilometers] away; everything in its path for miles downstream was destroyed [2:9-12].

In all, this tremendous landslide of 240 million cubic meters (more than 300 cubic yards) of rock claimed some 2,600 lives.

In the analysis following the disaster it was inevitably asked whether it could have been avoided. Most such disasters are considered natural, and no one is blamed. However, this was not the case here. The Italian government charged eight persons with negligence and manslaughter because despite the known geologic hazards, the dam had been constructed.

CONCLUSION

While landslides are a part of the natural sequence of the geologic cycle in which land is eroded and transported to form new land elsewhere, it is evident that many of the problems they cause are a direct result of human activities. The Vaiont

Disaster might have been avoided if the designers of the dam had considered the problems of slopes more carefully. The same conclusion can be applied to other slides where modifications of the pre-existing conditions lead to increased instability of surficial materials. Such an observation backs the arguments of environmentalists who call for rational planning in the use of sloping land. The idea of "slope zoning," which, like the zoning of floodplains would restrict the use of steeply sloping land, appears to be basically sound. However, it is doubtful whether such a plan will ever be widely implemented because of economic and social pressures for the use of slopes.

SUGGESTIONS FOR FURTHER WORK

1. Creep on slopes can be seen in many localities. Discuss the evidence of such surficial movement in the area in which you live. What types of slopes show the most movement?
2. How can landslides occur beneath the sea? When they do occur, how can the result influence people?

SELECTED SOURCES FOR FURTHER READING

Bloom, Arthur. Geomorphology: A Systematic Analysis of Late Cenozoic Landforms. Englewood Cliffs, N.J.: Prentice-Hall, 1978.

Chorley, Richard J., Stanley A. Schumm, and David Sugden. GEOMORPHOLOGY London: Methuen, 1984.

Ingmanson, D. G., and W. J. Wallace. Oceanology: An Introduction. Belmont, Calif.: Wadsworth, 1973.

Keefer, D. K. "Landslides Caused by Earthquakes", Geological Society of America Bulletin, Vol. 95, No. 4 (April), 1984.

McCarr, A., and R. C. Vorhis. "Seismic Seiches from the March 1964 Alaskan Earthquake."Washington: U.S. Geological Survey, Professional Paper, 544E, pp. E1-E43. 1968.

Schuster, Robert L. and Robert W. Fleming. "Socio-Economic Significance of Landslides and Mudflows", Proceedings of the Geologic and Hydrologic Training Program, Denver, 1984. Reston, Va.: U. S. AID Office of Foreign Disaster Assistance, and U.S.G.S., 1984.

REFERENCES

---

1. Keefer, David K., "Rock Avalanches Caused by Earthquakes: Source Characteristics", Science, Vol. 223, No. 4642 (March 23), 1984.
2. Kiersch, G. A., "Vaiont Reservoir Disaster," Geotimes, May-June, 1965.
3. Leighton, F. Beach, "Landslides", Geologic Hazards and Public Problems, Conference Proceedings. Santa Rosa, California: Office of Emergency Preparedness, Region 7, 1969.
4. Williams, K. The Snowy Torrents: Avalanche Accidents in the United States, 1967-1971, Washington: U.S. Department of Agriculture, Forest Service, 1975.

Part V
The Biosphere

Chapter 16

Energy Resources

The Context

In the late 18th century, muscle power — human and animal — provided the principal means by which wheels turned and people and goods moved from one place to another. Then, thanks to Mr. Newcomen's engine and a series of other inventions, there began a "new age," a time when more and more ways were found to substitute inanimate energy for muscle power. The result was a change so great as to be referred to as the "Industrial Revolution."

Wood and coal, burned to

release the solar energy stored long ages before, heated water to make steam to operate machinery. Later, in the last quarter of the 19th century, petroleum, another fossil fuel, began to replace coal and wood as a source of inanimate energy.

Today we are coming to realize that our world has a limited supply of these traditional sources of inanimate energy. As a result, we can recognize how essential energy is throughout our environment.

In fact, everything--an automobile, a suit of clothing, an apple pie, a TV program--involves an input of energy. This is true even for an intangible thing such as a college education, not only the proverbial midnight oil, but the energy that has gone into--and thus has been embodied in--a thousand other endeavors such as designing the buildings and writing the textbooks. In short, embodied energy confronts us wherever we turn. [5:25] Let us pursue this idea further.

Energy to Produce Energy

The energy in a barrel of oil becomes available only after a certain amount of energy has been expended in order to produce that oil. Those inputs include the energy required to mine the iron ore, process it into steel, fabricate the steel into drilling equipment, transport the equipment, and so on--even including the energy involved in training the engineers and others involved in the operation. If the amount of energy required to produce a barrel of oil is equal to or greater than the amount to be obtained from that barrel, there is, of course, little point in producing. It follows that the value of oil seeping from the earth's surface in Pennsylvania would be different from the value of oil that lies one and a half kilometers (a mile) below the ocean floor and 30 kilometers (twenty miles) offshore. Awareness of such basic considerations permits more accurate assessment of the world's energy resources and emphasizes the element of efficiency: energy inputs compared to outputs. The same principle appears in other parts of the energy scene, as when an automobile is evaluated: how many units of energy have to go into the machine in order to achieve the desired movement of the vehicle?

Improving Energy Efficiency

The unprecedented oil price increases of the 1970's encouraged efforts to reduce consumption of non-renewable energy resources by increasing efficiency as well as by reducing overall consumption and finding suitable alternative sources. This was especially true in the developed countries where improvement in efficiency has been dramatic. In less developed countries, however, conservation efforts have been less evident. By 1985 less developed countries were consuming nearly twice the oil used in 1973, whereas the United States, e.g., used no more energy in 1985 than it did twelve years before. All sectors of the American economy have been involved. Improvements in fuel efficiency of new cars mean that kilometers per liter (miles per gallon) are now more than 11 (26 miles per gallon), up by about ten (about 2.6 gallons) since 1973, and aircraft operating efficiency is reported to have increased by about 50% in the same period. Residential energy use declined by about 10% in spite of a 23% increase in the number of households. In the United States commercial sector better building designs and energy-efficient retro-fits contributed to a 35% decrease in overall energy consumption, and the technological potential exists for more progress. World-wide, for each 1% gain in energy efficiency, about one million barrels per day of petroleum equivalent could be saved. [9:96-98]

Petroleum's Expected Decline

At present, petroleum continues to supply nearly half the world's energy needs. For the United States alone this fuel accounts for well over 50% of the energy used. [9:2] Whether that oil comes from domestic or foreign sources depends upon a complex of economic and political factors, and whether oil supplies a greater or lesser proportion of the world's energy depends on its cost relative to the other sources. However, it seems likely that in the free world at least, while total energy consumption increases from about 205 quadrillion B.T.U.'s to about 250 quads, oil's share of total energy consumption is expected to decline slightly while all other major energy sources (i.e., gas, coal, nuclear power, hydroelectric power and wood waste) are expected to increase their overall share of total energy consumption. [9:A-3, A-15, B-4] [Figure 16-1].

ELECTRICITY FROM THE SUN

Among the oft-mentioned alternatives to conventional sources of energy, especially for generating electricity, is "solar", in this case meaning direct conversion. The amount of such energy reaching the earth is, "huge, variable, and diffuse". [4:172] Of course, at night or when the sun is otherwise obscured, either back-up power or storage batteries are needed. Moreover, energy in sunlight is of little value in this context until it has been concentrated, a process that to date remains expensive, whether measured in dollars and cents or in embodied units of energy.

Photovoltaic (Solar) Cells

We devote the rest of this chapter to a summary of the nature and promise of photovoltaic or solar cells. These cells, developed in the early 1950's, have the remarkable ability to concentrate solar energy as they convert it directly into electricity. Moreover, they rely on a clean and inexhaustible resource, and since they have no moving parts, maintenance costs can be virtually eliminated over a lifetime of 20 years or more. [See 3:50] Like transistors and semi-conductor integrated circuits, solar cells can be easily mass-produced, promising economies of scale.

How Do Solar Cells Work?

Solar cells vary, but most of them are derived from cylindrical silicon crystals, synthetically grown. The crystal cylinders are meticulously cut into thin slices (in thickness, 300 micrometers or less, equivalent to about three pages of this book) [2:90] which can be treated to produce two types of silicon: "n-type", with negative (n) charge carriers (electrons) and "p-type", with positive (p) charge carriers or "holes" (deficiency of electrons). Fusion of an n-type slice with a p-type slice produces a wafer containing an n-type to p-

Figure 16-1. Free World Energy Consumption Shares.
Source: U.S. Department of Energy, <u>Energy Security: A Report to the President of the United States</u>, Washington, 1987, p. A-4.

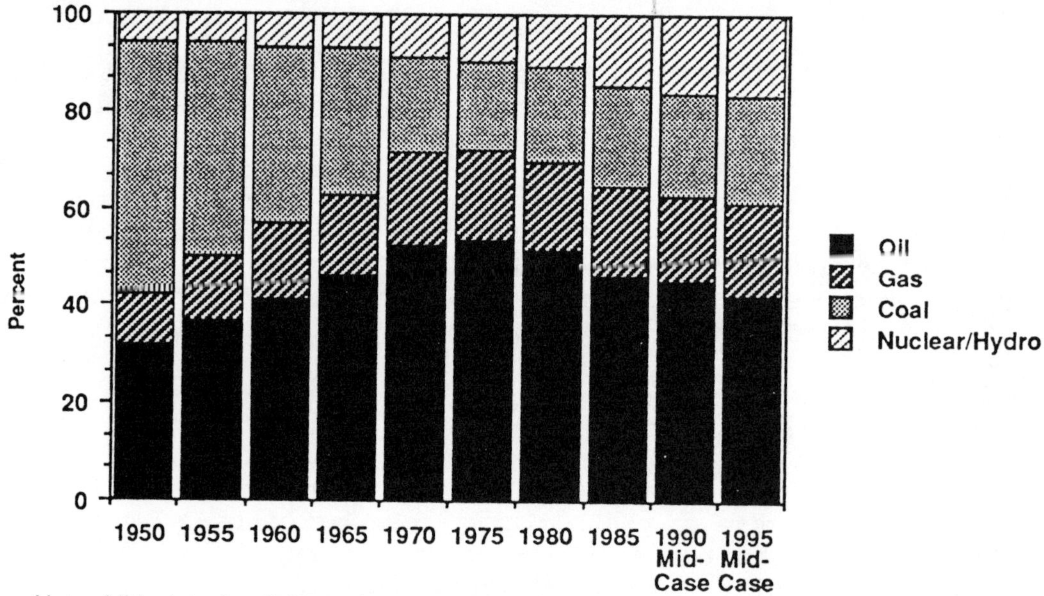

Note: Midpoints for 1990 and 1995 were derived from range of uncertainty between lower and higher oil price cases.

type junction and a built-in electric field. Incoming light (photons) can dislodge electrons which then flow across the junction to the "p-type" layer, setting up a flow of electrons which can be collected as an electric current. (Figure 16-2) The amount of current varies with the light intensity and the surface area of the cell. Cells are responsive to more than one wavelength, and they can produce energy whenever there is light, even on overcast days or in ordinary room light. [See 7] To compensate for periods of darkness they can store energy in batteries, and with the use of inverters they can supply alternating current for uses that require it.

Figure 16-2. How Solar Cells Make Electricity. Source: Solarex Corporation, Building Blocks of Solar Electricity, 1979. (A brochure).

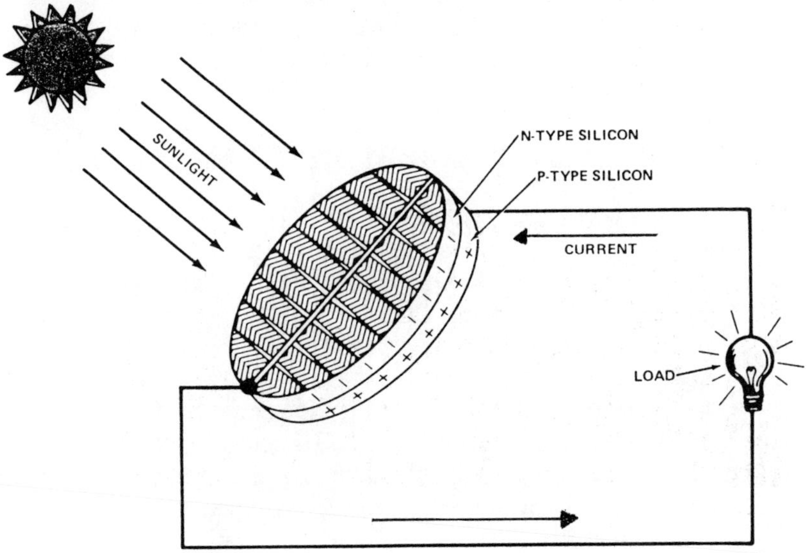

"Energy flow within a crystalline-silicon solar cell utilizes, in principle, about 44% of the solar energy to which the cell is sensitive. About 16% of the energy is lost by various processes within the cell, leaving a theoretical limit of efficiency of 28%. Further losses reduce the actual efficiency of the cell in converting sunlight into electricity to between 14 and 22%." [2:90] Efficiencies of this magnitude have only recently been achieved for mass-produced cells, thanks to rapidly improving technology.

"It is often pointed out that the conversion efficiency of a solar cell is less than half that of generating systems, based on nuclear power or fossil fuel. The comparison is nonsensical. At a conversion efficiency of 38% a steam turbine fueled by oil is wasting 62% of the expensive fuel. (Indeed, the fuel came from sunlight-associated processes far less efficient than a solar cell.) Moreover, the generated energy may cause thermal pollution, acid rain and atmospheric contamination. On the other hand, even if a solar-cell installation is only 10% efficient in converting sunlight into electricity, it is effectively utilizing an otherwise neglected energy source that costs nothing." [2:89-90]

Declining Costs of Sun-Generated Electricity

High cost is the prime factor that has delayed the building of large generating plants using solar cells. Nonetheless, as measured in dollars, relative cost of solar electricity has dropped sharply over the past decade. The cost of a solar cell module in 1974 was $50 per watt of peak power. Consequently, the cost of electricity produced by such a module would have been extremely high--$3 per kilowatt hour at a time when electricity generated by fossil fuels and nuclear energy was from five to ten cents per kilowatt hour. This difference in cost has inspired a vigorous research effort to reduce the cost of photovoltaic electricity. New materials, different designs, improved concentrators to focus more light on individual cells, better mass-production methods, and lively competition among manufacturers have all contributed to bringing the cost of a module down to about $5 per peak watt. "Further reductions, perhaps by another order of magnitude to less than forty cents per peak watt for thin-film cells such as amorphous silicon, can be expected in the next decade." [2:91]

Practical Uses for Solar Cells

The desirable qualities of solar cells have long been recognized, and they have proved themselves in a number of applications, among them the supplying of power for hand calculators and for earth satellites. They provide the power for Coast Guard buoys, and private and commercial boat owners use solar cells to keep batteries charged and to power lights and other marine equipment. Traffic signals in the United States, educational television facilities in remote schools in Africa, weather and stream-gaging equipment in the California mountains and water well pumps in Bolivia are but a few of many other examples. [See 10]

A Million-Watt Generating Facility

A pioneer among the photovoltaic power-generating systems--and the world's largest--is the one-megawatt (i.e., one million watt) installation at Hesperia, near Victorville, in desert California. Built on an eight-hectare (twenty-acre) site owned by a utility company, since its completion in 1982, it has been supplying to that company's distribution system enough solar-cell-generated electricity to service 400 homes. [11:10; 1:3; see also 6 and 8]

The mass-produced solar cells at Hesperia are grouped into modules, and the modules are fixed into 108 sixteen-foot-long arrays, each containing 256 modules. [11:10; 1:3] The arrays are mounted a meter or so (a few feet) above the ground, facing the sun, and arranged in neat rows across the site. (Figure 16-3) Each array is mounted to permit computer-controlled tracking; that is, continuous positioning to follow the diurnal and seasonal movement of the sun, and thus to maximize the intensity of insolation striking the cells.

The Hesperia plant's 1985 performance was reviewed in an independent report which showed that when 1000 watts of sun energy were being received per square meter (about 1.2 square yards) of surface area, and when the ambient air temperature was 20 degrees C (about 64°F), the plant was rated at 728.2 kilowatts of direct current with a conversion efficiency of 7.2%. In the same year, the plant produced more than 2,244 megawatt hours of direct current--96% of its energy potential. Almost all of the energy losses were caused by high winds and resultant tracking problems. "The Hesperia photovoltaic power plant was a reliable energy producer during 1985." [8:Abstract]

Rising Commercial Production of Photovoltaic Cells

To meet the growing demand for solar cells world-wide a

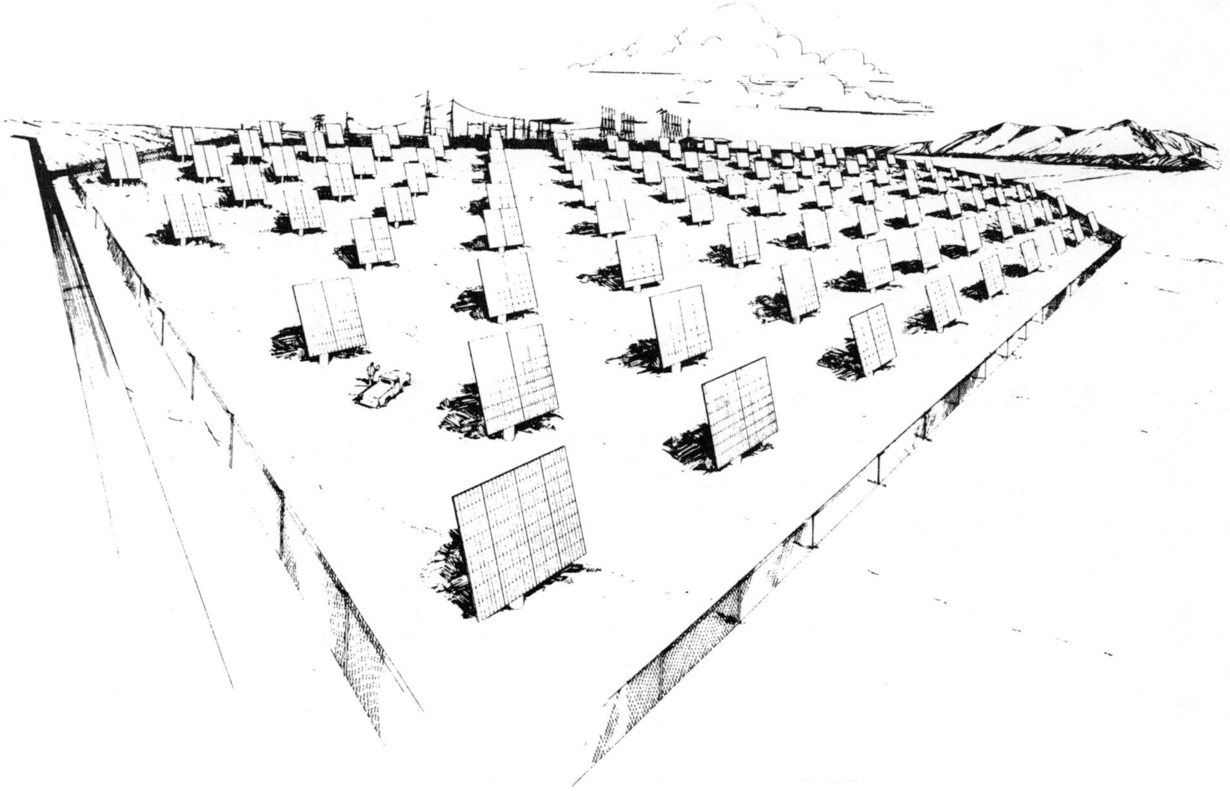

Figure 16-3. Artist's drawing of the photovoltaic power plant near Hesperia, San Bernadino County, California. Solar cells are grouped together in modules which automatically track the moving sun. This plant provides enough power to meet the electrical needs of 300 to 400 typical homes. (Courtesy Siemens Solar Industries.).

number of manufacturing firms have entered the field. Value of annual production as of 1982 was estimated at about $100 million --three times what it had been in 1980--and in 1986 the figure had climbed to $380 million. Generating capacity doubled in 1982, reaching 9.2 million watts--still, a minuscule fraction of the world's total electrical generation from all sources. Even so, the total capacity of solar cells shipped in 1985 amounted to 20.5 megawatts, and it reached 24.7 megawatts the next year. This increase occurred in spite of reduced government-and-tax-sheltered purchases; it indicates a strengthening of the commercial photovoltaic market. Japanese manufacturers, concentrating on very small units for consumer goods such as calculators and watches, now lead the world in total capacity produced. American manufacturers rank second in total production, but they dominate the photovoltaic power module market. [12: Summary] European manufacturers place nearly equal emphasis on large and small systems. They supply photovoltaic systems to many third world countries for village electrification and water pumping, and they also supply traditional markets for commercial equipment and remote residences. [12:1] Besides Japan, the United States and Europe, small but significant commercial production is now coming from firms in India, Brazil, China, Saudi Arabia and Canada. [12:15]

CONCLUSION

Among the developing alternatives to fossil fuels is the photovoltaic cell. It converts sun energy directly into electrical energy, and further research and development effort should bring greater efficiency to this process; that is, higher outputs, not only per unit of solar energy incident <u>on</u> the cell, but higher outputs of energy per unit of embodied energy <u>in</u> the cell itself. Lower monetary costs per unit of solar power are expected to follow, and additional commercial photovoltaic plants will be built. [See, e.g., 2:92] Through the next few decades such systems, along with other alternative energy sources, may well contribute to extending the life of the world's diminishing reserves of fossil fuels and to making the un-failing, non-polluting, inexhaustible sun an even more important energy resource. Perhaps one of these alternative sources will usher in an age as revolutionary as the one which began with Mr. Newcomen's engine in England two centuries ago.

SUGGESTIONS FOR FURTHER WORK

1. Review the first two paragraphs in this chapter and then list as many examples as you can of energy embodied in a college education. (If you need further help on this, see Reference 5, below.)

2. Assume you had $1,000 available to invest in an energy-related industry. (e.g., petroleum, coal, solar, geothermal, etc.) Which one would you choose? Why?

SELECTED SOURCES FOR FURTHER READING

Brown, Stuart F. "Chasing Sunraycer Across Australia", Popular Science, Vol. 232, No. 2 (February, 1988), pp. 64 ff.

Fri, Robert W. "New Directions for Oil Policy", Environment, Vol. 29, No. 5 (June, 1987), p. 16 ff.

Hamakawa, Yoshihiro. "Photovoltaic Power", Scientific American, Vol. 256, (April, 1987), pp. 86-92.

"Harnessing the Sun: Solar Energy Programs", Sandia Technology, Vol. 11, No. 1 (April, 1987), pp. 46 ff.

Landsberg, Hans H. "Rethinking Energy Security: the Case for Coal in the United States", Environment, Vol. 29, (July/August, 1987), pp. 25 ff.

"New Solar Cell: Competition for Coal?", Environment, Vol. 28 (June, 1986), pp. 23-24.

Sawyer, Stephen W. Renewable Energy: Progress, Prospects, State College, Pennsylvania: Commercial Printing, Inc., 1986.

REFERENCES

1. "ARCO Solar and Southern California Edison Announce Construction of One Megawatt Photovoltaic Facility", ARCO Solar News, Vol. 2, No. 2 (April, 1982), pp. 2-3.

2. Hamakawa, Yoshihiro. "Photovoltaic Power", Scientific American, Vol. 256 (April, 1987), pp. 86-92.

3. "Harnessing the Sun: Solar Energy Programs", Sandia Technology, Vol. 11, No. 1 (April, 1987), pp. 46 ff.

4. Kash, Don E. and Robert W. Rycroft. U.S. Energy Policy: Crisis and Complacency, Norman: University of Oklahoma Press, 1984.

5. Odum, Howard T. and Elizabeth C. Odum. Energy Basis for Man and Nature, New York: McGraw-Hill, 1976.

6. Rhodes, G.W. ARCO One-Megawatt Photovoltaic Power Plant, Albuquerque: B.D.M. Corporation, 1984.

7. Solarex Corporation. Building Blocks for Solar Electricity. (Brochure) (1979).

8. Stokes, K.W. Hesperia Photovoltaic Power Plant: 1985 Performance Assessment: Final Report, Palo Alto, California: Electric Power Research Institute, 1987.

9. U.S. Department of Energy. Energy Security: A Report to the President of the U.S., Washington: U.S. Department of Energy, 1987.

10. "Usable Electricity from the Sun", Cleveland: Technology Utilization Office, Lewis Research Center, (no date).

11. "Wait Till the Sun Shines", Forbes, Vol. 133, No. 10 (May 7, 1984).

12. Watts, R.L. and S.A. Smith. Photovoltaic Industry Progress from 1980 through 1986, Washington: U.S. Department of Energy, 1987.

Note: The foregoing chapter appears in the "Biosphere" section as a reminder of the basic link between plant and animal life on the one hand and fossil fuels on the other. Thus, directly or indirectly, the sun is responsible for the world's muscle power as well as its fossil fuels and most other sources of inanimate energy. Even so, since this chapter emphasizes the harnessing of sunlight directly, it could have been placed appropriately elsewhere in this book of readings.

Chapter 17

Amazonia Development:
At What Cost?

The Context

A quick glance at a world population map will show that population is very unevenly distributed. There are vast regions where population density is very low. Most of these regions are polar lands, deserts, or mountainous areas where climate and landforms are less suited for human habitation. There is, however, one area of the earth that has a low population density that does not fit this general scheme. It is the equatorial rainforest of South America.

The image most of us have of the Amazon rainforest is one of an incredible luxuriance of plant growth inhabited by exotic animals and people. It has an aura of mystery and is surrounded by many preconceptions and a great deal of misinformation. After an initial appraisal of the forest, we might concur with the views of early colonists who assumed that here, where natural vegetation grew thick and fast, was a region that could be incredibly productive when used for agriculture. Outsiders are often inclined to see the rainforest as an area of vast riches awaiting discovery, exploitation, and profit. Such a conclusion is far from the truth, for in the forests there are environmental constraints that place severe limitations on development. Despite the best intentions of the planners, development schemes to open the interior of Amazonia have experienced serious setbacks caused by these limitations. In this chapter, we shall discuss some of the problems experienced by those who have settled in the Amazon basin.

BACKGROUND

Many observers feel that Brazilians have been more interested in fast speculative gain than in the rational development of the nation's resources. Thus, much of Brazil's history has been characterized by economic booms and busts. Three major cycles dominated much of Brazil's economic history. The initial cycle, from the mid-16th to the end of the 17th century, was based on the introduction of sugar cane in the northeast and gave rise to the first significant permanent settlement. The second major cycle, beginning at the end of the 17th century, was based on gold and diamonds in Minas Gerais and led to the first significant settlement away from the coast. The last cycle, beginning in the mid-19th century, was dominated by coffee and focused on Sao Paulo. In each of these cycles there was a spectacular rise in the production and value of a commercial product which had a rapidly expanding market. A period of great profit followed, but eventually the market declined as other locales in the world were able to produce the products and successfully compete in the world marketplace. About the turn of the 20th century, a short-lived rubber boom flourished in the Amazon Basin, and at other times commercial products such as Brazil wood, cacao, and oranges played minor roles in Brazil's economic history.

During the time these cyclical economic booms and busts were taking place, a large segment of the Brazilian population continued to depend on subsistence agriculture. Agricultural development occurred mainly on lands once covered by semi-deciduous forests; but by the mid-20th century, very little virgin forest remained, and soil erosion had become a pervasive problem. Throughout most of Brazil's history, the vast interior remained a remote hinterland and except for the short-lived rubber boom was largely ignored in terms of economic development.

Now the situation has changed dramatically. During the past three decades, the government has taken major initiatives to transform the emptiness of the backlands. The interior has been seen as an enormous storehouse of resources, but the size of the area and the difficulty of access had, in the past, protected it from major exploitation. In 1960, a new capital city, Brasilia, was created to stimulate development of the interior. Shortly afterward, Brazilian President Kubitschek inaugurated a 2,175 kilometer (1,350 mile) dirt road from the new capital to the Amazon port of Belem. A major turning point came in 1970 when the government faced staggering social and economic problems. Prompted by a severe drought in the northeast which drove millions from their land, officials announced a hastily prepared plan to build a network of long-distance roads across the interior. The core of this proposal was a highway stretching from the Atlantic coast to the frontier of Peru.

The government saw the opening of the rainforest as a means of bolstering the economy through a resurgence in agriculture and other resource development. In addition, settlement projects would relieve the population pressure in the Northeast. However, scant attention was given to the problems that might be encountered in the Amazonian environment.

Construction of the main east-west highway was undertaken

just three months after the plan was announced in September, 1970; and subsequently the road network has been gradually expanded, opening up large areas of the forest for development (Figure 17-1).

Figure 17-1. The Trans-Amazon Highway and its location relative to other features in Brazil's Amazonia.

The planners' strategy was to bulldoze a path 3,200 kilometers (2,000 miles) long straight through the heart of the Amazonian rainforest and build a dirt road rather grandly named the Trans-Amazon Highway. Officials believed that in the following decade one million people could be moved into the region and settled on 100-hectare (250-acre) farms that would be carved out along the highway margins (Figure 17-2).

Although the government was prepared for the sophisticated engineering skills and hard work such a road would require, the rainforest presented unexpected obstacles. In a fascinating

Figure 17-2. The Trans-Amazon Highway. (Photo by Peter L. Gould from Keystone Press Agency).

article about constructing a road through "Brazil's Green Hell", Hummerstone wrote:

> These are the days of heat and dust, of rain and mud, days of malaria and dysentery, of stinging ants, mosquitoes and flies, in one of the earth's most hostile regions.... There are also days of furious struggles with snakes, jaguars, and hostile Indians [3:16].

Such an account describes only the beginning of the problems, for camps were often so remote that food and supplies had to be dropped to them by air, heavy equipment got stuck in the mud, and the remoteness of the construction site placed dire hardships on the construction workers.

Although conditions for the workers were bad, the project brought even worse miseries for the Indian tribes. With the road has come social disruption and disease, and today many experts doubt that the indigenous cultures can survive the settlement of the Amazon.

FAILED DEVELOPMENT SCHEMES

Before examining actual problems that have resulted from the Amazonian Highway development, we shall look at two grandiose projects that have taken place in the Amazon. These projects, which were ambitious in their design and well capitalized, have been dramatic in their failure. Henry Ford, in 1927, and Daniel Ludwig, in 1967, sought development in the forest to meet commodity needs on the world market. Both men had reason to believe in the economies of scale and were successful entrepreneurs. But both men underestimated the difficulties of Amazonian development and ultimately had to accept costly failure.

After World War I, the growing use of the automobile significantly increased the demand for rubber. At that time, rubber production was monopolized by the east Indian producers; so the Ford Motor Company, under Henry Ford's direction, decided to establish its own plantations in the home of the rubber tree, Brazil. In cooperation with the Brazilian government, Ford selected a site on the Tapajos River, and the community of Fordlandia was established.

Almost immediately, technical problems beset the operation. The initial site was hilly, and intense tropical rains quickly eroded the soils and flooded the bottoms of valleys and ravines. A substantial part of the property was rocky and unsuitable for trees, exportable timber used for quick cash was rapidly depleted, and an invasion of South American leaf blight attacked many trees.

A number of environmental factors led to Ford's difficulties. In this vast tropical realm there is a great deal of local and regional diversity in soil, flora, and fauna, but for human exploitation, the most important distinction is the difference between the soils of the Varzea and the Terra Firme. The Varzea, which are limited to some 2-4% of the Amazon basin, consist of flood plain soils that have supported a settled population since prehistoric times. Like the Nile Valley prior to the construction of the Aswan High Dam (see Chapter 9), the annual flooding renews and fertilizes these soils, making them suitable for continuous intensive agriculture. On the other hand, the Terra Firme, or the extensive uplands, are soils extremely poor in nutrients. Here, the natural vegetation of the climax tropical forests effectively recycles nutrients essential to plant growth. But when the forest is eliminated, the replenishment of nutrients is lost. In addition, increased soil erosion and leaching contribute to depleted soils and to a rapid decline in harvest productivity.

Ford's choice of a site on the Terra Firme, therefore, represented a grave error. This error was compounded by the removal of vegetative cover. Natural vegetation can exist on the infertile soils only by maintaining a rapid nutrient cycle. In fact, one of the great paradoxes of the rainforests of the equatorial realm is that the profuse vegetation exists on some of the world's worst soils. The vegetation thrives because of the ecological balance between the amount of nutrients produced by plants and the rate at which the plants reuse them. Nutrients, derived from the litter of the forest, are rapidly taken up by other plants with a minimum loss from the system.

Fordlandia offers a classic example of the pitfalls in the management of monocultures in the tropics. The operation was plagued by plant disease and other environmental problems throughout its existence. The difficulty of attracting workers from other regions to the remote jungle outpost and competition in the form of cheap synthetic rubber ultimately convinced the Ford Motor Company that the plantation would never produce rubber economically. In 1945, with over fifteen million U.S. dollars invested, Ford sold the plantation as a cost cutting measure to the Brazilian government for five hundred thousand dollars [4].

This investment loss was small compared to that of shipping magnate Daniel Ludwig several decades later. Like Henry Ford, Ludwig was a pioneer determined to triumph over the tropical rainforests. In 1967, Ludwig purchased three million acres in the northern part of the state of Para where he established Jari. The Jari project was intended to respond to an envisioned world-wide paper shortage. By the early 1980's, Jari was the largest development enterprise ever attempted by a private individual. The project, which included the world's largest tree farm, comprised some 250,000 acres of tree plantations, 2,600 miles of dirt roads, 45 miles of railroad, a Kaolin mine, a pulp complex, rice farms, and cattle and buffalo operations. Some 30,000 workers lived and worked in the area surrounding Monte Dourado, a planned settlement built from scratch.

Jari, like Fordlandia, suffered technical setbacks from the origination of the project. A rapid growing East Indian tree Gmelina had been selected as the species to be grown; however, ignoring foresters' advice, laborers used bulldozers to clear the natural forests for planting. The heavy machines scraped off most of the thin topsoil and so compacted the soil that most of the seedlings died. Subsequently, the plantation managers were forced to turn to manual forest clearing. Lack of preliminary site evaluation and knowledge of the environment of the region proved costly. Gmelina did much less well in Brazil than it had in other geographic regions. The tree was poorly adapted to sandy soils, and ultimately was replaced by Caribbean pine and eucalyptus on these sites.

Like Fordlandia, Jari suffered technical problems and a mis-evaluation of the fragile Amazonian ecosystem. In 1982, rising costs, labor shortages, and other difficulties led to the liquidation of the development. A government-backed Brazilian consortium purchased the project for a total of 280 million dollars. It was estimated that this gigantic project cost Ludwig about three quarters of a billion dollars. These development schemes demonstrate the potential problems and the need for cautious, careful assessment of the Amazonian environment.

CONTEMPORARY AGRICULTURAL DEVELOPMENT PROBLEMS

In recent years, the relationship between the Amazonian environment and its people has reached a turning point. The newly constructed roads created during the past two decades have opened the frontier to exploitation as people come in search of land and livelihood. Government sponsored colonization projects and large scale investment in hydroelectric facilities and mining have resulted in a flood of immigrants to the region. In recent years, over 200,000 new settlers have arrived annually.

The migrants represent diverse groups, both with respect to homeland and experience. Pioneers include affluent growers and ranchers, subsistence farmers from all over the country, and a large number of people with diverse occupational backgrounds. The vast majority, however, come from the ranks of the dispossessed who have entered a land that is unfamiliar. These people have chosen to move to the frontier to escape the extreme poverty of their home areas where inequitable distribution of land has been coupled with the pressure of high population growth. Much of the farm land in the east, for example, stands idle because owners are more interested in speculative profits from rising land values than in producing crops. In addition, a small segment of landowners controls a disproportionate amount of the arable land. In 1975, fewer than 1% of the land owners controlled 43% of the land in properties that were greater than 1,000 hectares (2,470 acres). Fifty-two percent of the landholders occupied 3% of the land area in holdings of under 10 hectares (about 25 acres) [1:409]. To seek the opportunity to own their own land, farmers have moved west to Amazonia.

PROBLEMS OF COLONISTS' FARMING SYSTEMS

Pioneer farmers, entering the Amazon, employ agricultural systems with many essential differences from the traditional agricultural resource management utilized by the Amerindians and Caboclo populations in the past. These differences adversely affect the fragile Amazonian ecosystem. It is these environmental issues that we want to consider here. As we have discussed in the Fordlandia and Jari sections, the forest is maintained within a closed nutrient cycle. How easily that system can be disrupted and destroyed by agriculture is illustrated in Figure 17-3. Changing land uses modify the surface soil condition and increase leakage of nutrients from the soil. Under natural conditions, shown in Figure 17-3a, the surface is protected from heavy rains, and the various plant species, many with different nutrient requirements, rapidly recapture the nutrients made available as the forest litter decomposes.

As practiced by the indigenous forest people, the slash and burn cultivation (swidden), shown in Figure 17-3b, involves clearing the forests, allowing the debris to dry until it can be burned, cropping the area for a few years, and then abandoning the area to forest regrowth. These traditional farmers have adapted to local conditions over a long period of time. They usually utilize intercropping which is characterized by a high degree of crop diversification and the layering of crops with adequate fallowing. The soil is minimally disturbed and recovery is rapid. Traditional swidden systems, which are sustainable in areas of low population density where a given site has a sufficient recovery period, are well adapted to the tropical rainforest environment.

Many of the new colonists, who are now the dominant tropical forest farmers, find adaptation to the new environment difficult, and use inappropriate techniques. Typically, what occurs is something like this. As the major highways and their branches are cut through the forest, settlers follow and move in to clear areas for farming. Crops such as maize and upland rice are planted; however, within two to three years the heavy rains leach the soil nutrients, and surface erosion begins, Figure 17-3c. Because the rapid decline in soil fertility makes the land unsuitable for cash crops, pasture grasses are planted, and the land is sold for cattle ranching. The farmer then moves on to another area, deforests more land for planting, and the process repeats itself. There are, of course, many variances to this basic process.

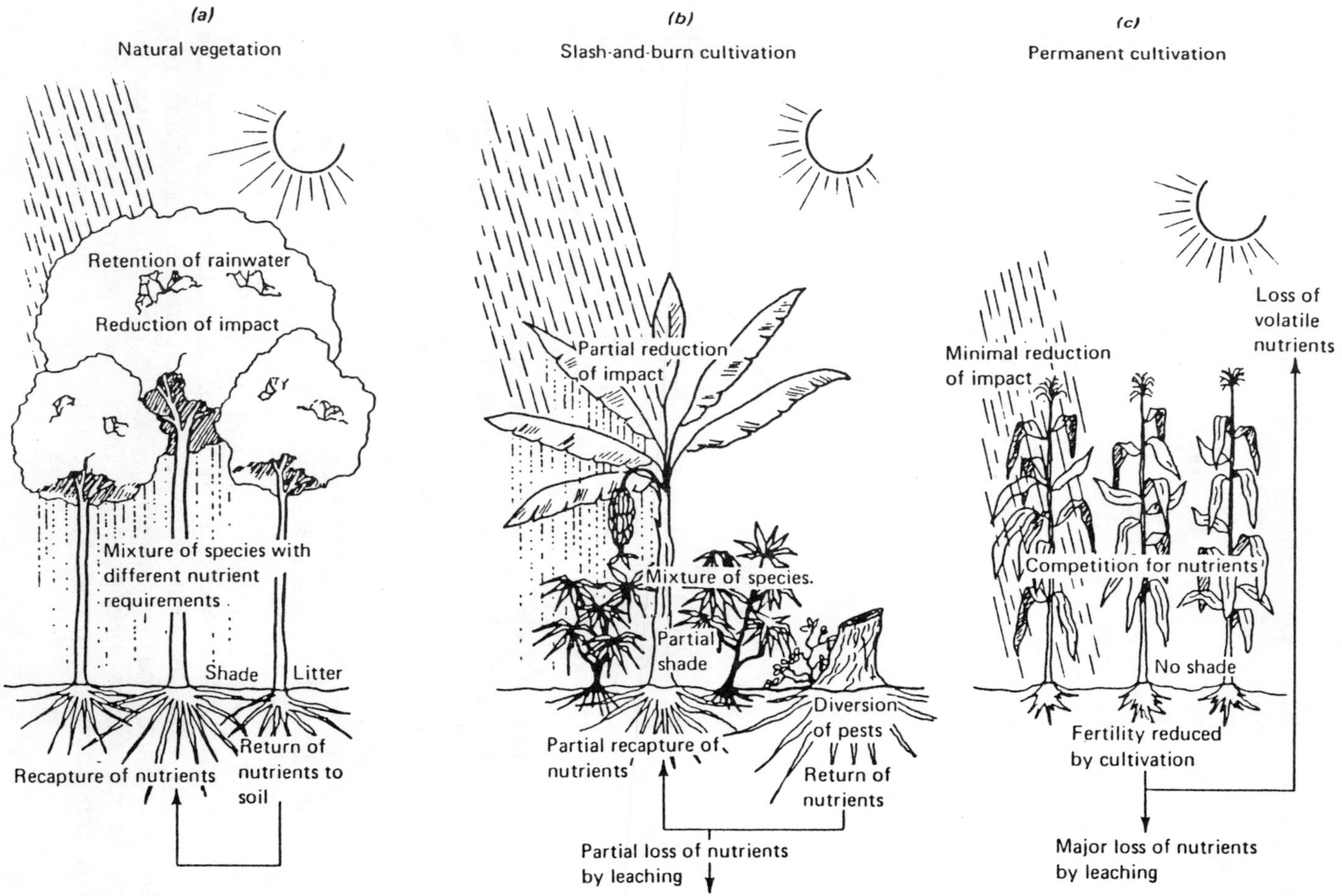

Figure 17-3. The effects of different kinds of land use on equatorial lands. (From B.J. Meggers, Amazonia: Man and Culture in a Counterfeit Paradise (Arlington Heights, IL: AHM Publishing Corporation, 1971), p. 24.

Other factors have contributed to colonists' difficulties. Because these pioneers frequently make poor site choices and often practice monoculture, they farm large, homogeneous fields which are more susceptible to pests and disease problems. Typically they crop too long and either have too short a fallow period or fallow crops with pasture. These practices lead to a rapid depletion of the soil, and therefore the system is unsustainable. Many other disadvantages are present in simplified agricultural ecosystems. Dickenson has nicely summarized some of the major differences between monoculture and high diversity agricultural systems in the tropics (Table 17-1).

During the 1960's and the 1970's a significant proportion of the development focused on clearing pasture for cattle ranches. Although hard data are lacking, the predominant land use change is the clearing of forests for pasture. As with farms, the initial nutrient enrichment from burning of the trees is at the expense of total available nutrient stocks, and the pasture productivity declines rapidly in subsequent years. Cattle also trample the soil until it becomes compacted and hard. Because the ability of the soil to absorb water is reduced, pasture quality is further diminished. Declining soil fertility, overgrazing, and weed and pest problems have led to the abandonment of degraded pastures after only a few years of use.

As a result of these factors, beef productivity has been low, and many of the projects have been abandoned. Although there is no longer a government push for new cattle projects, the subsidies and economic incentives are still available for ranching projects.

Despite the problems experienced by the cattle ranchers, individuals continue to take advantage of government incentives. The clearing of land is a legally defined improvement and is often the first step in establishing ownership rights. The production of beef is frequently a minor consideration for the land owner. The conversion of land to pasture is important for social and institutional reasons. Cattle and land continue to symbolize gentlemanly prestige. Land has great speculative value, and the lure of speculative profit has continued to

Table 17-1. A Comparison of Monocultural and High-Diversity Cropping Systems in the Tropics. Source: J.C. Dickinson, "Alternatives to Monoculture in the Humid Tropics, <u>The Professional Geographer</u>, Vol. 24 (August, 1972), p. 219.

Characteristics	Modern Monoculture	High-Diversity Systems
1. Net Production	High--given heavy fossil fuel subsidy.	Moderate--but sustain-able indefinitely.
2. Biochemical Contribution To Diet	Low--complete diet dependent upon assembly of food from spatially dispersed monocultures at considerable cost.	High--complete energy vitamin, and protein from a variety of cultigens and animals at one site.
3. Species Diversity	Low--generally devoted to a single variety or breed.	High--both in total and number of individuals of each species.
4. Space Utilization	Poor--single strata frequently with bare earth unoccupied by photosynthetic material.	Excellent--three-dimensional space largely filled by plants adapted to direct and diffuse light.
5. Inherent Stability	Low--dependent upon the fossil fuel economy for nutrients and chemical and mechanical control of outbreaks of host-specific pests and diseases.	High--competitive exclusion of weeds by diverse food plants, avoidance of pest epidemics through diversity of host plants. Analogous to natural systems.

6. Nutrient Cycles	Open--large proportion of all nutrients applied to crops is lost seasonally to leaching.	Tight--minerals lost by early successional annuals taken up by perennial crop plants. Detritus cycle well developed.
7. Economic Stability	"Boom or Bust"--with optimal environmental and market conditions and large fossil fuel investments, high yields, and profits are possible. Vulnerable to environmental stress and market fluctuations beyond the farmer's control. Labor requirements highly seasonal. Tendency for mechanization to replace human labor.	High--variety of food produced for region or national consumption assures a market for some crops. Flexibility to switch plant energy flow from direct marketing to increased animal production is practical. Low capital investment makes subsistence on a quality diet feasible. Yield can be programmed throughout the year, as can labor input.
8. Social Viability	Volatile--economies of scale tend to concentrate management decisions, production and profit in the control of a national socioeconomic elite of foreigners. Foreign control of markets is common,	Initially volatile, but contributing to long-term stability. Emphasis on the direct involvement of peasant farmers in ecological and economically viable systems of quality food pro-

and dependence on fossil fuel inputs creates need for a satellite relationship with developed countries.	duction. Functional systems would be a valuable resource in serious agrarian reform. Low ratio of land-owning managers to agricultural laborers.

channel land use decisions in the direction of pasture rather than other more intensive agricultural activities. Susanna Hecht has concluded that cattle ranching in the Amazonian region is demonstratively unprofitable. Ranches exist only because of generous government tax breaks, subsidized credits, and the revenues that come from land speculation [2]. On natural grassland or thinly forested areas, cattle ranching is an appropriate land use. However, in the tropical forests this land use is of dubious sustainability. Schmink notes:

> One calculation found that the production of enough ground beef for one hamburger requires the destruction of approximately seventy five kilograms (two hundreds pounds) of living matter, including some 20-30 different plant species, perhaps 100 insect species, and dozens of bird, mammal, and reptile species. ...While one hamburger has an easily measurable market value, the one-tenth ton of forest required to produce it does not [6:169].

DEFORESTATION AND ITS EFFECTS

Another, although somewhat more speculative, result of the destruction of the tropical forest is related to the potential effects on climate. Locally, there is no question that deforestation causes alteration in relative humidity, solar radiation received at the ground level, and soil and air temperature. At the regional scale, deforestation results in changes in hydrologic balance and increased stream flow. Recent studies suggest that deforestation also may cause a significant decrease in regional rainfall. Globally, forests play a significant role in regulating the CO_2 concentration in the atmosphere. Extensive tropical forest burning converts the forest biomass into carbon dioxide and may increase the intensity of the greenhouse effect which results in the warming of the earth's atmosphere. This contribution to atmospheric carbon dioxide, while significant, is far less than the very large conversion of fossil fuels to carbon dioxide. Global climatic changes are extremely complex, and what is needed today is basic research supported by hard data to document the impact of tropical deforestation.

Perhaps the most significant problem related to tropical deforestation is species extinction. Approximately 1.7 million species have been identified on the earth. The tropical forests contain millions of species of plants and animals that are found nowhere else on earth. Because tropical forests are so poorly known and biologically rich, most scientists accept the statement that the forests contain species numbering at least 5 million, with many researchers believing the total could exceed 10 million. Predictions of species loss based on recent land use trends suggest that some 15-20% of all species on earth are likely to be extinct by the year 2000 [7:86]. Most of this loss will occur in the tropical forests. Many of these species have not been studied scientifically, and extinction would eliminate them before we have a chance to learn about them, or in many cases even know of their existence. The quality of our lives will be diminished, not only because of the loss of the aesthetic beauty intrinsic in these diverse forms, but also because of their importance in the earth's evolutionary heritage. If we squander this heritage, we will know nothing of the potential uses of these species.

Wild species serve human needs in many ways. For example, more than half of modern medicines come from the natural world, including such tropical forest products as cuarie, quinine, and strychnine. Other useful products include lumber, ornamental plants, spices, nuts, fruits, and an array of industrial products including volatile oils, resins, gums, waxes, and dyes.

Forest wildlife is also an important resource. Wildlife provides a sizeable fraction of the diet of many tropical forest people. For all of us, wildlife provides important psychological, cultural, and spiritual values. In addition, tourists come to view wildlife and to engage in wildlife-based recreation. Money now spent on these activities suggests that wildlife tourism has significant economic importance.

CONCLUSION

In this chapter we have used the Brazilian Amazon to illustrate some of the pitfalls of tropical rainforest development. For the Brazilian government, the Amazon has been a place to settle many landless peasants, to tap vast natural resources, and to secure national borders against real or imagined threats from neighboring countries. Much of the development has been directed at solving problems that have occurred outside the Amazon region. Among these problems are the need for development of export earnings to lessen the national debt, the alleviation of rural and urban poverty in the eastern part of the country, and the provision of investment opportunities for businessmen and speculators from more capital-rich areas. Human occupation of the Amazon, however, has often led to devastation, violent conflicts over land titles, and harm to the indigenous cultures. Development on a sustainable basis is a difficult and poorly understood challenge.

Deforestation is a problem that extends well beyond Brazil, and is today one of the world's most pressing land use issues. Tropical forests covering some 6% of the earth's surface are being cleared at the rate of 20-40 hectares (50-100 acres) per minute. This world-wide loss of trees has had varied effects. The loss of forests means the loss of soil fertility, increasing soil erosion, and degradation of fragile lands. It means changes in the water supply, increased run-off, reduced navigability of rivers, and added pressure on fisheries. It means a decrease in fuelwood, fodder, and timber. It means a loss of species and biodiversity, and it means the loss of indigenous people, their culture, and their knowledge of forest use.

Policymakers, and scientists who act as advisors, must find methods of developing tropical forests that offer sustainable livelihoods for the people of these countries. Unfortunately, far too many of the world's tropical countries are engaged in short-term, short-sighted development ventures, and their governments are unwilling to make politically unpopular decisions to control forest devastation.

It is necessary that we recognize the problems of tropical rainforest development. The path forward, however, is in the development of techniques and systems of sustainable resource utilization which will enhance and preserve our environmental inheritance. Investments in research and land use planning, in population control, and in systems that are usable on a long-term sustainable basis can preserve this unique biotic heritage.

SUGGESTIONS FOR FURTHER WORK

1. Investigate some of the major arguments for and against wilderness preservation.
2. How can the United States press for tropical forest preservation when many people who live in the countries where these forests are located question the past forest removal in our country? Look for basic similarities and differences between tropical and mid-latitude forests.
3. Many of the rich food-producing countries of the world have destroyed their forests as a means of gaining wealth. Are ecologists from these countries being hypocritical when they urge people from tropical areas to preserve their forests? Defend your point of view.
4. In class, debate the following: Wealthy nations should/should not help tropical countries bear the financial costs of preserving their forests.

5. How well adapted is (a) hunting and gathering and (b) slash-and-burn agriculture to the ecology of the equatorial rainforest?

SELECTED SOURCES FOR FURTHER READING

Burley, F.W., and P. Hazelwood. "Tropical Forest Action Plan" Journal 86, The Annual Report of the World Resources Institute. Washington, D.C., 1986.

Institute of Ecology. "Man in the Living Environment", Report of the Workshop of Global Problems. Madison: University of Wisconsin Press, 1972.

Longman, K.A., and J. Jenik. Tropical Forest and Its Environment. Longman Scientific and Technical, 2nd edition, 1987.

Meggers, B.J. "Amazonia: Man and Culture in a Counterfeit Paradise. Chicago: Aldine-Atherton, 1972.

REFERENCES

1. Fearnside, Phillip M. "Agriculture in Amazonia." Key Environments--Amazonia. G.T. Prance and T.E. Lovejoy, eds. Oxford: Pergamon, (1985), pp. 393-418.
2. Hecht, S.B. "The Environmental Effect of Cattle Development in the Amazon Basin." Center for Latin American Studies, University of Florida.
3. Hummerstone, R.G. "Cutting a Road Through Brazil's 'Green Hell,'" New York Times Magazine, (March 5, 1972), p. 16.
4. Janik, J. et al., Plant Science, San Francisco: W.A. Freeman, (1969), pp. 476-79.
5. Rankin, Judy McKean. "Forestry in the Brazilian Amazon." Key Environments--Amazonia. G.T. Prance and T.E. Lovejoy, editors. Oxford: Pergamon, (1985), pp. 369-392.

6. Schmink, Marianne. "Big Business in the Amazon," People of the Tropical Rain Forest. Julie Sloan Denslow and Christine Padoch, eds. Berkeley: University of California Press, (1988), pp. 163-174.
7. World Resources Institute and the International Institute for Environment and Development, World Resources, 1986. New York: Basic Books, 1986.

Chapter 18

Environmental Future

The Context

Our environmental situation continues to decline faster than ever. This is demonstrated by the growing spread of pollution of both the water and atmospheric systems, erosion of topsoil, tropical deforestation, desertification, mass extinction of species, depletion of the ozone layer, and climatic dislocations. Compounding the impact of these problems is growth in human numbers in developing countries, together with the associated factor of extensive poverty. In this chapter we ask why our environmental situation should continue to decline, despite a remarkable outburst of awareness on the part of the general public and its political leaders since the time of the 1972 Stockholm Conference on the Human Environment.

At the time of the Stockholm Conference on the Human Environment in 1972, fewer than ten governments had environmental agencies. Today the total is around 140. In other ways too the past fifteen years have seen a remarkable outburst of awareness of environmental issues, on the part of both the general public and its political leaders. Yet our environmental situation has continued to decline. Why should this be so? And what should we do about it -- that is, how differently should we try to do a better job with our environment?

This chapter tackles these two questions. Moreover it broaches them within a context of interdependence among the community of nations -- not only economic but ecological interdependence. This factor of interdependence is of salient significance to the environmental cause, presenting both problems and opportunities.

OUR ENVIRONMENTAL PREDICAMENT

A World of Accelerating Change

We are not only imposing ever-more profound impacts on our biosphere, we are doing so at ever-increasing rates. We have been injecting pollutants into the environment at such progressive speed during the past few decades that certain parts of the pollutant burden would leave their mark for many more

----

[1] This chapter has been adapted from an editorial by Norman Myers, "Emergent Aspects of Environment: A Creative Challenge," The Environmentalist, Volume 7, Number 3, (1987), pp. 163-74. Used with permission. Professor Myers lives in Oxford, United Kingdom. He is a consultant in environment and development and has published widely on these subjects.

decades to come even if we were to halt the process forthwith. Most of the recent spread of deserts cannot be repaired for centuries. Tropical deforestation has worked up such momentum within the past few decades that the damage can surely not be made good for at least a millennium (Figure 18-1). Climatic dislocations from recent human activities may well persist for thousands of years. The current extinction spasm will leave our natural world so impoverished that evolution will not restore the biotic injury in less than millions of years. We are imposing all these mega-scale changes on our planetary ecosystem within the space of a single century, at most, the twinkling of a geologic eye.

There are also superscale changes afoot in our economic and political situation. The population explosion has not only added more people to the human community since 1950 than in all the previous history of mankind; the main phase of population growth is still to come during the course of the next few decades. Associated with the population phenomenon, there is an unprecedented amount of absolute poverty in the Third World.

This means we must seek, during just the next few decades, to build several new human worlds on top of the one we have already built -- bearing in mind that the environmental foundations, and hence the economic basis, of the present one are all too shaky. To cater for twice as many people, we shall need three times as much in the way of food, fiber and buildings if we are to catch up with quality as well as quantity of need. We must also generate three to four times as much energy.

The economic and environmental background serves to introduce an equally significant change: the growing interdependence that characterizes our global community. We are learning that joint problems warrant joint answers, and that we can best become prosperous by becoming prosperous together. In short, we face a situation of one form of mega-change superimposed on another, and another, each reinforcing the rest. We also face a fast-growing gap -- a gap between our capacity to impose changes on the world we live in, and our capacity to manage our basically changed world. Far from closing that gap, we are

Figure 18-1. Tropical deforestation has worked up such momentum that damage will surely take at least a thousand years to repair. Erosion on outskirts of Dumoga National Park, Indonesia. (Photo: WWF/Michèle Dépraz. Copyright, World Wildlife Fund, Gland, Switzerland. Used by permission).

watching it grow bigger. We still have time to close the gap: but only just enough time.

All this means that as the global economy becomes ever-more integrated, and as we make ever-greater demands on our natural-resource base, we face two major sets of institutional challenges. The first is with respect to our national systems of government, and the second our modes of international cooperation. We shall consider the first of these in this opening section, the second later in the chapter.

Integration of Environment and Economics

A prime task is to integrate two sets of values, those of environment and of economics. When we make use of man-made assets, such as equipment and buildings, we write off our use as depreciation. But we have not learned to evaluate our environment as productive capital, even though we utilize it as such. When we cut down forests, over-work croplands until the soil erodes, and utilize our skies as a free garbage can and our rivers as cost-less sewers, our income as revealed through gross national produce (GNP) indices actually registers an increase. Yet we eventually have to pay, often more heavily than if we had acknowledged the cost in the first place.

An obvious illustration is the case of the acid rain. In Europe, some 200,000 sq km of forest, or an area almost the size of Britain, are injured if not dying. [12] In the worst-hit country, West Germany, the costs to the timber industry are conservatively estimated at $1 billion a year, and the overall costs to the economy at $2.8 billion a year. In addition, the European Community believes that acid-rain damage to Community-wide forests, crops, lake fisheries, buildings and human health amounts to at least $13 billion a year [6].

The acid-rain problem is spreading, widely and fast. There are signs of it in China, Malaysia and Brazil -- even of an acid haze over the Arctic. There could hardly be more vivid evidence that we need to integrate the values of environment and economics, until a unified system of accounting reflects the world outside the window, a seamless world that does not recognize man-made divisions of reckoning.

For an instance of the economic penalties of environmental mismanagement in a developing country, consider Ethiopia. The country's forests, which covered 25 percent of the country as recently as 1940, have declined to only three percent today. The result has been widespread soil erosion from the traditional farming areas in the country's highlands. It can be roughly reckoned that this loss of soil with its plant nutrients cuts Ethiopia's agricultural output by at least one million tons of food per year -- equivalent to two-thirds of all relief food shipped to the country in 1985. Moreover, as trees disappear and sources of fuelwood go too, people take to burning cattle dung and crop residues. So much of these materials are now being used as fuel instead of fertilizer that there is a further agricultural output foregone, worth some $600 million a year, or no less than 30 percent of recorded value added in the entire agricultural sector in Ethiopia. To restore tree cover and to safeguard topsoil, would, according to a United Nations estimate, have cost some $50 million a year if undertaken in due time during the 1960s. Yet in 1985 the outside world spent almost $500 million on relief food alone for Ethiopia.

We can trace a further linkage between deforestation and development, this one with respect to the energy sector. Among the best prospects of plentiful and renewable energy in the Third World is hydropower. Of all electricity now consumed, 38 percent comes from this source; and the proportion could grow a good deal higher by the end of the century -- provided water supplies remain adequate. But a recent World Bank survey [17] of 200 major dams built in the past 45 years reveals that sedimentation, caused by washoff of soil in the wake of deforestation, leads to an average of two percent loss of storage capacity every year. Even a one percent loss would mean reduced output of electricity on such a scale that, were it to be generated by burning oil at a price of $15 a barrel, would cost $3.5 billion in the single year of 2000 -- or a sum roughly equivalent to one-tenth of all the aid now being supplied annually by the European-based Organization for Economic Cooperation and Development (OECD) and the Organization of Petroleum Exporting Countries (OPEC).

An End to Business As Usual

This clash between our system of economic accounting and environmental management points to a basic conclusion. The future will not be a simple extension of the past. If we try to pursue a path of business as usual, we shall encounter some altogether unusual consequences. In essence, we are consuming our environmental capital rather than the income from the capital.

We must beware, then, of those who assert that we can somehow 'get by' until we have restored full economic growth, whereupon we shall possess the discretionary funds to deal with environmental problems. This view suggests that environmental problems are sideline affairs that can wait until we have the luxury of wealth and to spare. But if economic growth is based on misuse and overuse of environmental capital (as is too often the case), then we shall find we are undercutting the very capacity of our economies to keep on growing. The question to ask about environmental protection is not 'Can we afford to do it eventually?' It is, 'Can we afford not to do it immediately?'

Some Policy Responses at National Level

In face of these radical problems, what sort of radical response shall we envisage? We have tried environmental impact assessments, technological fixes, and a host of other such measures. Helpful as they are, they are plainly not enough, since our environmental situation is deteriorating faster than ever. So we need to take a long look at our environmental policies -- or rather the lack of them.

A notable example is agriculture, and especially with respect to certain policies of developed nations. While pursuing an immediate goal of stimulating productivity and profitability, the policies throw up long-term problems in that they over-reward farmers for over-production of food at cost of environmental values [3]. For instance, they encourage farmers to over-apply pesticides, with the environmental contamination that is becoming pervasive. They prompt farmers to over-apply nitrogenous fertilizers, to the extent that water bodies in several European countries now feature nitrates above the health limits set by the European Economic Community (EEC) and the World Health Organization (WHO). They lead farmers to make careless use of scarce water for irrigation. They induce farmers to open up new land by clearing woodlands and forests (Figure 18-2), even at the loss of environmental services such as watershed functions. In the United States soil erosion imposes on-site costs of $2.5 billion a year, and off-site costs of $6 billion [15].

As for more distant parts of the world, there is a plethora of policy inducements to fell tropical forests. In Central America, if a man owns a patch of forest and does not do anything with it, he is taxed because he fails to 'improve' it. If he decides to log it, clear-cut it or slash-and-burn it, he is deemed to have served a socially useful function, so he gets a tax break. In Indonesia there are so many hidden subsidies to foster over-cutting of rainforests that whereas a conventional accounting asserts that timber-exports add $2 billion to GNP each year, a more rigorous appraisal [13] reveals they impose a net loss of $1.6 billion. Overlogging is a problem in many tropical forest countries. Export revenues are now worth more than $7 billion a year, yet they are projected to fall to only $2 billion by the year 2000, when a majority of present timber exporters are expected to become net importers. As for non-logging types of exploitation, cattle ranchers and small-scale farmers in Brazil, Columbia, Peru and Ecuador are actively encouraged through fiscal and tax incentives to establish enterprises in Amazonia, at cost to the rainforests (Figure 18-3).

Especially in the field of energy we find a host of concealed subsidies that emphasize the role of fossil fuels in modern societies, for instance by offering support for extraction of fossil fuels and by offering tax-and-trade support for large vehicles [14]. Conversely there is ample scope to foster greater efficiency of use -- which not only has obvious advantages for national economies, but assists the international need to curb atmospheric pollution, whether in the form of acid

Figure 18-2. Human settlement in sensitive tropical ecosystems is often induced by government schemes of support. (Photo: WWF/Russell A. Mittermeier. Copyright: World Wildlife Fund, Gland, Switzerland. Used by permission).

Figure 18-3. Road built through Amazon Basin tropical rainforest helps to destroy rich ecological habitat. (Photo: WWF/Michael Freeman. Copyright: World Wildlife Fund, Gland, Switzerland. Used by permission).

rain or carbon dioxide buildup. We could make much more use of the energy source that is cheaper, more widely available and less exploited than any other, and that is, conservation worth perhaps 25 percent of all energy consumed [4]. In Britain alone the savings could amount to £3.3 billion (about $4 billion) a year, or almost three times more than 1986 revenues from North Sea oil; while in developed nations as a whole, the total could reach $94 billion a year, or three-quarters as much as the value of all commercial energy consumed in the Third World each year.

All this -- the growing complexity of economic and ecological interactions, plus the need to apply a 'pre-fit' approach to projects -- takes place in a world that is becoming increasingly interdependent. This key factor of interdependence serves to aggravate the growing number of environmental problems that are no longer national in scope. Fortunately it also offers opportunity -- splendidly expanded opportunity -- for their resolution. We shall now look at the phenomenon of interdependence, and as it manifests itself in several forms.

INTERDEPENDENCE

Economic Interdependence

Since the early 1950s, the global economy has steadily become more internationalized through an almost continuous process of liberalization of the international economic order. Between 1950 and 1980, the volume of trade increased almost six-fold [11]. Moreover economic interdependence has been expanding not only through trade but through commerce, investment and other workings of the marketplace. These economic linkages have been fostered through parallel forms of interdependence, notably in the shape of communications, transport, and labor migration. Today we find the global economy is becoming yet more interlinked, and not only through the integrative consequences of expanding trade and investment, but through trans-frontier inflation, global recession and debt-servicing difficulties.

As a measure of the importance of these economic linkages,

economists estimate [16] that a three percent increase in economic growth in the developed world serves to increase economic growth in the developing world by one percent per year. If in 1980 the developed-world economies had grown at the rate they maintained between 1965 and 1975, Third World economies would have grown by an extra $25 billion -- thus expanding potential markets for developed-world exports.

Ecological Interdependence

Economic interdependence is increasingly matched by ecological interdependence, which likewise reflects on many issues of economic well-being. Pollution, which was once basically a national concern (Lake Erie, the River Thames, Minemata Bay), is now often a regional if not a global affair, as witness the spread of acid across much of Europe and North America, and the wind-borne dispersal of radioactive material from Chernobyl. Tropical deforestation in one country can impact on several other countries, as witness the damage caused by Himalayan deforestation to the 1.25 million sq km Ganges River system with its 350 million people in Nepal, India and Bangladesh. Desertification in one country readily becomes a process that spills over into other countries, as witness the Sahel and Southern Africa. Soil erosion, while intrinsically local in immediate impact, impinges on the soil-based capacity of the entire world to feed itself: the loss of topsoil in both Ethiopia and Iowa deserves to be a matter of concern to people everywhere. Extinction of species and elimination of genetic reservoirs, especially in the Third World tropics which are biotically richer than all the rest of the world, diminishes the amount of genetic diversity required to maintain the productivity of major crops throughout the world, and especially those, such as maize and wheat, of developed nations.

Particularly at global level, and with respect to atmosphere and climate, the build-up of carbon dioxide, due to combustion of fossil fuels and burning of tropical forests, together with increase of other 'greenhouse' gases, are expected to bring warmer temperatures to all nations, but more rainfall to some and drought conditions to others. Depletion of the ozone layer reduces crop yields and fosters skin cancer, not only in the 'culprit' nations but throughout the world [7:18].

In Third World instances there is a further link-up, this one with population growth. When population grows ultra-rapidly, beyond the planning capabilities of government and beyond the carrying capacities of environmental resources, there is a major threat to economic advancement. One manifestation, and stemming from decline in carrying capacity of the natural-resource base, is the phenomenon of 'environmental refugees', as witness the mass movements of destitute people within Africa. The environmental problems of Haiti formerly concerned Haiti alone, but now, through migratory throngs of Haitians fleeing their devastated landscapes, they become a domestic problem for the United States in the form of costly resettlement programs for refugees.

Economic-Ecological Linkages

A further form of interdependence arises through economic-ecological linkages at international level. A salient instance is the 'hamburger connection', whereby North Americans' demand for cheap and non-inflationary beef has promoted the spread of cattle ranching in Central America, the pasture lands being established almost entirely at the expense of tropical forests [9]. Forest cover in the region has declined from around two-thirds in 1950 to only one third or so today; at present rates of depletion there will be virtually none left by 1995 or shortly thereafter. The consequences in terms of soil erosion, disruption of water flows down rivers, and siltation of hydropower facilities, are widespread and severe [5].

Several further instances arise. For instance, the cassava connection between the European Community and Thailand, fostering deforestation; cash-crop exports to Europe as a source of desertification in the Sahel; the beef trade between Botswana and the European Community, promoting desertification; and the developing world/developed world trade in tropical hardwoods.

In each of these instances of economic-ecological linkages, there is a common outcome. The end-product consumer serves to foster environmental degradation in lands far removed from his own. He does it all too unwittingly, but he does it effectively and increasingly.

The most pervasive instance of economic-ecological linkages lies with the relationship between poverty and environmental degradation in the Third World. The most numerous agent of environmental degradation is the impoverished peasant, who over harvests fuelwood stocks, over grazes grasslands, causes topsoil to erode, stimulates the spread of deserts, and engages in unsustainable slash-and-burn farming of forestlands. Through a variety of economic forces, which range from inequitable land-tenure systems and lop-sided patterns of development, to international patterns of trade and aid, this man finds himself relegated to the fringes of the development process. His marginal status drives him to seek his living in marginal environments, i.e, those that are too wet (tropical forests), too dry (arid lands) or too steep (montane environments) for sustainable human settlement -- these being precisely the environments that are most susceptible to degradation. While recognizing that his actions of today are destroying his means of livelihood tomorrow, he sees no alternative because of the pressures of sheer poverty.

The main, and often the only, way to relieve this man's destructive impact is to enable him to share more solidly in the benefits of development -- notably by practicing intensive rather than extensive agriculture, also stabilized rather than migratory (shifting) agriculture. To achieve this there is need for many parties -- governments in both developing and developed nations, international aid agencies, development banks, and commercial investors and other business interests -- to make common cause in order to promote a basically essential goal, that of sustainable economic growth throughout the Third World.

Interdependence: A New Era of International Relations

This brief review of interdependence -- both economic and ecological interdependence -- reveals that it is becoming more difficult for nations to retain their traditional self-image as autonomous entities that exercise sovereignty in conventional sense, viz., sole and exclusive sovereignty. The unitary nature of the planetary ecosystem in which all nations interact makes is apparent that no nation can shield itself from environmental degradation in other nations. Nor can it consider that responsibility for its actions generally ends at its own frontiers. Substantial environmental impacts can be transmitted far and wide through water systems and the atmosphere -- just as, by virtue of the increasingly integrated nature of the global economy, they can be transmitted through (and indeed stimulated by) trade and investment systems.

In summation, interdependence is a dominant theme of our times, becoming ever-more significant. It is much more than interconnectedness, since it reflects dynamic interactions that are intrinsically two-way in operation. The economic linkages have hitherto been the more influential, as well as the more readily recognizable. But they are now being supplemented if not supplanted in many cases, by ecological determinants--which in turn are becoming the new 'objective reality' of international relations.

All this postulates an imperative of greater international cooperation. We live not only in interlinked economic and political systems, but our shared biosphere leaves us commonly vulnerable. As a consequence, it is virtually impossible for any one nation to protect its national interests through unilateral action. Alternatively stated, we find that certain 'national' interests are becoming co-extensive with the interests of other nations, often of many other nations -- in some circumstances, with those of the entire community of nations. In short, a number of conventionally perceived 'national' interests are becoming strongly convergent with an array of global interests. This new situation offers great scope for creative initiatives on the part of political leaders and policy makers.

In the main, however, interdependence still tends, in certain quarters at least, to be covert rather than acknowledged. Not all political leaders perceive its basic character and extent, and not all are ready to articulate its implications through institutional initiatives of proper scope. In other words, interdependence still works its way, in many instances, by default rather than by design -- as witness the limited decision-making capacity of an economic institution, OECD, and an environmental institution, The United Nations Environment Program (UNEP), among other attempts at international cooperation.

Hence the rationale for political leaders to come up with formalized expressions of interdependence. That is, there is growing incentive for them to act, rather than react, in response to fundamental shifts in the international arena. At present there is a growing asymmetry between issues and responses: interdependence tends to be perceived as a problem to be countered, rather than an opportunity to be grasped. True, a number of political leaders are starting to recognize the issues and their import. But too many political leaders still give more attention to the obvious costs of action than to the concealed costs of inaction. They thus fail to develop the institutional capacity to confront the challenge. Meanwhile the issues grow bigger and more numerous, faster -- even though interdependence is not something that we shall need to address at some stage in the future. Rather it is an established fact right now, something we need to manage to our advantage forthwith. Let us remember, moreover, that the problems of interdependence will either be solved through deliberate response, or they will be resolved through unwitting reaction-- to the benefit or detriment of society at large.

POLICY IMPLICATIONS: A CONCEPTUAL FRAMEWORK

Plainly the new era of interdependence represents a basic departure in international relations generally, and with respect to environment in particular. An array of issues that have traditionally been treated as primarily domestic -- forestry, agriculture, water stocks, pollution, population -- are coming to be seen as issues with growing international ramifications. To this extent, they deserve to be broached through cooperative action among the international community than has been the case hitherto.

Let us now review some specific measures.

1. Recognition of Opportunity

First and foremost, there is need for political leaders to take explicit cognizance of the nature and scope of interdependence and its importance for environmental issues. They will thus acknowledge the new 'objective reality' -- which need not be perceived as a constrictive constraint on established patterns of activity, rather it can be seen as a creative incentive to constructive initiative.

2. Expanded Potential for Policy Changes

We no longer enjoy the option of a world without interdependence. We face only the option of ways to live with interdependence, i.e. to manage it. This postulates a quantum advance in international cooperation, to reflect the qualitative shift in international relations. An approach of 'the same as before, only more so and better so' will no longer suffice. An unprecedented degree of cooperation is becoming an imperative. Equally to the point, it is becoming an opportunity too, for expanded policy initiatives that reflect an environmental situation where we shall all win together or we shall all lose together.

3. New Perceptions of the Nation-State System

Arising from the above is the need to acknowledge that our affairs still tend to be run for the most part by intrinsically national institutions that derive from the nation-state system -- a system that is traditionally competitive rather than cooperative. Fortunately, a growing number of nations are starting to swap certain images of independence for the benefits of interdependence, at least in the economic arena (albeit in rather laissez-faire manner). But in the environmental arena, the process remains largely unplanned, undirected, uncoordinated--unmanaged in general.

While we face an expansion of interdependence among nations, this need not imply any diminution in the role of the nation-state. Rather it will require a basic shift in mode of operations -- often with an increase in the 'positive profile' of the nation-state. Collective action of the sort envisaged need not downplay sovereignty. On the contrary, it should serve to emphasize sovereignty as a functional rather than an idealized concept. As an established administrative unit, the nation-state will surely find its role enhanced as it undertakes the task of 'thinking globally while acting locally'.

4. Need for New Institutions

To reiterate a central point, the decision-making nexus, still centered on national governments, is more and more limited for decision-making on international concerns. Alternatively stated, certain of our institutions are not mature enough to meet the new challenges. This is hardly surprising when we remember that they were not designed for international cooperation of scope and scale to match up with today's demands.

5. Costs of Action - and of Inaction

Finally, the question of costs. To undertake environmental measures of sufficient scope, the costs will be substantial, totalling tens of billions of dollars per year. But to put them in perspective, we should consider the concealed costs of inaction. We have noted that in 1980 desertification was imposing costs on commodities affected to an extent of at least $26 billion a year -- a figure that throws light on the budget for the Anti-Desertification Action Plan of $4.8 billion a year. In the Sahel, measures to meet food needs during 1984-87 were expected to cost a total of $1,390 million. Yet a concurrent four-year effort to stem further desertification (reforestation, anti-soil erosion measures, etc.) was scheduled to cost only $108 million -- and if undertaken 10 years ago, the effort would have markedly reduced the scale of the recent famine-relief operations [19].

So, too, with fuelwood deficits in Nepal, where diversion of cattle dung and crop residues from farmlands to domestic hearths means a decline in food-growing capacity. By the year 2000 this will lead to food production foregone on a scale to match Nepal's entire crop output in 1970 [10].

By far the greatest costs of inaction appear to lie with delay in developing energy policies to cater for threats of a greenhouse-affected world. This applies especially to our inefficient use of fossil fuels, which generate carbon dioxide among other greenhouse gases. Suppose we were to exploit, immediately and fully, the various forms of energy-efficient technologies available to us already; and suppose we were to supplement them with measures to internalize the non-greenhouse effects of coal and shale-oil burning (effects that include acid rain among other forms of pollution). According to this scenario, and if we assume that our reliance on fossil fuels will continue roughly as at present, we find that an aggressive commitment to energy efficiency could hold power-plant emissions of carbon dioxide to 7 billion tons a year worldwide; but if efficiency levels remain in levels of the mid-1970s, annual emissions would rise to 17 billion tons by the year 2050 [2].

What would be some costs associated with a greenhouse affected world? In the field of irrigation alone, we find that plans for new installations of dams, canal networks and the like, projected to cost $200 billion by the year 2000, are posited on the assumption that climatic conditions will not change significantly. In a planet with higher temperatures (let alone with shifts in precipitation patterns), many of these installations will prove pointless; and many more will be needed to compensate for climatic changes. The cost for adaptive measures in just 15 percent of irrigated lands could bear as much as a further $200 billion [1]. When we add in the need for extra flood-control measures, new crop varieties and other agricultural responses, the total could soar to $450 billion.

Regrettably the momentum that has generated energy-efficiency gains of as much as 2 percent per year during recent years is now threatened by what may be termed the 'third oil crisis'. The return of cheap oil means we could quickly lose our recent momentum in efforts to increase energy efficiency. If we were to sustain the rate of increase over the next half century (and there is plenty of reason why we could, without harming economic growth rates), we could halve carbon dioxide emissions worldwide [8].

Unfortunately we do not work well with fifty-year time horizons. We tackle mere ten-year problems with five-year plans, staffed by two-year personnel working with one-year budgets. What counts as 'urgent' is what lies right under our eyes. Yet if the urgent drives out the important today, the important can become the imperative tomorrow -- when we shall surely have far less scope to confront it.

CONCLUSION

The thoughts presented in this chapter are not an idealist vision. Rather they are the acceptance of scientific facts. From this premise, a crucial two-fold challenge arises. First, how to shift from a largely scientific understanding to a political perception? And second, how to express the new concepts via policy measures and institutional initiatives? We have plenty of mechanisms to uphold and further the interests of nation-states. We have all too few to promote the interests of nation-states working together, in a community where collective welfare is often greater than the sum of the parts. Yet what is perceived today as a desirable but 'impractical' prospect will come to be seen tomorrow as a political imperative -- though by tomorrow, our room for maneuver may be severely constrained if we remain enmeshed in conventional strategies that reflect a by-gone world.

Let us not underestimate, of course, the size of the challenge ahead. It will represent one of the greatest changes ever to confront nations since the emergence of the nation-state as an established concept. It will probably prove to be a greater change than those precipitated in the past by widespread wars. Let us also remember, however, that the change is already in the making, albeit very little and very late. Nations already cede slivers of sovereignty to international bodies such as the United Nations and the Organization for Economic Cooperation and Development; they cede significant slices of sovereignty to regional groupings such as the North Atlantic Treaty Organization and the Association of Southeast Asian

Nations, and to trade organizations such as OPEC; and they cede considerable chunks of sovereignty to integrative units such as the European Economic Community. So a start has been made.

This enables us to assert that the new approaches, radical as they may appear, are not only desirable but feasible as well. Who would have thought at the time of the Stockholm Conference in 1972 that a Mediterranean Action Plan would become practicable, involving several parties who are long-standing antagonists (Israel and Syria, Greece and Turkey, Egypt and Libya, etc.)? How much more difficult and costly would an Action Plan be to formulate today? As we face the daunting challenges ahead, we can take heart from the ground-breaking precedent of this endeavor in international cooperation that works. The key component is this: pressing environmental needs have served to foster a sense of creative interdependence among the Mediterranean nations. Economic well-being and political stability have come to be recognized as indivisible -- as indivisible as is the common environment.

To this extent, environmentalism helps to articulate the shared interests of nations: confrontation gives way to cooperation. By promoting a consciousness of the biosphere's unity, as of society's unity, the environmental cause may even encourage governments to adopt a more constructive approach toward problems of other sorts -- notably the manifold questions of sustainable development within a safe-guarded biosphere.

REFERENCES

1. Brown, L.R. and seven others. State of the World. New York: W.W Norton, 1987.
2. Cheng, H.C., et al. Effects of Energy Technology on Global CO_2 Emissions. Washington: U.S. Department of Energy, 1986.
3. Conservation Foundation. Agriculture and the Environment. Washington: The Conservation Foundation, 1986.

4. International Energy Agency. Energy Conservation in IEA Countries. Paris: Organization for Economic Cooperation and Development, 1987.
5. Leonard, J. Natural Resources and Economic Development in Central America: A Regional Profile. Rutgers: TransAction Press, 1987.
6. Mandelbaum, P. Acid Rain--Economic Assessment. New York: Plenum Press, 1985.
7. Miller, A.S. and I.M. Mintzer, The Sky is the Limit: Strategies for Protecting the Ozone Layer. Washington: World Resources Institute, 1986.
8. Mintzer, I.M. A Matter of Degrees: The Potential for Controlling the Greenhouse Effect. Washington: World Resources Institute, 1987.
9. Myers, N. "The Hamburger Connection: How Central America's Forests Become North America's Hamburgers," Ambio, 10 (1), (1981), pp. 3-8.
10. _________. "Environmental Repercussions of Deforestation in the Himalayas," Journal of World Forest Resource Management, 2, (1986), pp. 63-72.
11. Organization for Economic Cooperation and Development. Economic Outlook. Paris: Organization for Economic Cooperation and Development, 1983.
12. Pearce, F. Acid Rain: What It Is, and What Is It Doing To Us? Harmondsworth, Middlesex, UK: Penguin Books, 1987.
13. Repetto, R. and M. Gillis, (eds.). Public Policy and the Misuse of Forest Resources. New York: Cambridge University Press, 1987.
14. _________ and M. Kosmo, (1986). Government Subsidies and Energy Mismanagement. Washington: World Resources Institute, 1986.
15. Runge, C.F. (ed.). The Future of the North American Granary. Ames: Iowa State University Press, 1986.
16. Sewell, J.W., R.E. Feinberg, and V. Kallab, (eds). U.S. Foreign Policy and the Third World: Agenda 1985-86. Washington: Overseas Development Council, 1985.
17. Sfeir-Younis, A. Soil Conservation in Developing Countries. Washington: The World Bank, 1986.

18. Titus, J. (ed.). Effects of Changes in Stratospheric Ozone and Global Climate. Washington: Environmental Protection Agency, 1986.
19. World Bank. World Development Report. Washington: The World Bank, 1985.

INDEX